This cookbook was born from the creative hearts and minds of the folks here at the Chia Café Collective. Like all of you, we love good ingredients, good recipes, and good company.

Authors

Barbara Drake, Daniel McCarthy, Deborah Small, Leslie Mouriquand, Cindi Moar Alvitre, Craig Torres, Abe Sanchez, Lorene Sisquoc, Heidi Lucero, Tima Lotah Link

© 2018 by the Chia Café Collective
All rights reserved. At no time may any part of this book be reproduced or transmitted by any means or in any form, electronic or mechanical, including photocopying, recording, or by any information storage or retrieval system, without prior written permission from the authors, except for the inclusion of brief quotations such as in a review.

Distributed by Heyday
P.O. Box 9145, Berkeley, California 94709
(510) 549-3564
heydaybooks.com

Designed by Tima Lotah Link
All photographs © Deborah Small

Printed on demand by Lightning Source, USA

Library of Congress Control Number: 2017930449

10 9 8 7 6 5 4 3

"*Cooking the Native Way* is not just about ingredients or culinary techniques, it's about a whole astonishing way of being, in community and in communion with the many plants, creatures, and spirits who help to bring us our daily bread, tortillas, or acorn mush."

— *Gary Nabhan*, Ph.D.

Director, **Center for Regional Food Systems**

"This book is a must-have for every cooking enthusiast on the planet."

— *Lois Ellen Frank*, Ph.D.

Chef/Owner, **Red Mesa Cuisine**

"Marinated in the indigenous resilience and wisdom of the Native peoples of southern California and seasoned with creativity and kinship, I can't wait to share this treasure with other communities, students, and foodies!"

— *Melissa Nelson*, Ph.D.

Editor, *Original Instructions: Indigenous Teachings for a Sustainable Future*

CONTENTS

1 | **THE CHIA CAFÉ COLLECTIVE**
Who we are, where we started, and what we believe.

7 | **A DEDICATION TO THE "PRESERVING OUR HERITAGE" AND THE "MOTHER EARTH CLAN"**
Honoring two organizations who have lent us strength, guidance, and knowledge over the years.

13 | **NEVER TAKE MORE THAN YOU NEED**
A history of natural resources depletion across the southern California landscape.

19 | **NATIVE FOODS AND WHY THEY MATTER**

23 | **PLACES**
Chia Café Collective cooking adventures throughout southern California.

24 | **IDYLLWILD**
Lessons on the vast edible and medicinal bounty that surround us all.

37 | **CATALINA ISLAND**
Camping under the stars, cooking on the cliffs, making memories and mesquite tortillas.

42 | **TORRES MARTINEZ**
Rabbit is WAY better than chicken!

49 | **MALKI MUSUEM**
Coming together for the annual Agave Harvest & Roast.

54 | **SHERMAN INDIAN HIGH SCHOOL**
Bringing Native American culinary secrets back to life at an Indian boarding school.

59 | RECIPES
Our cooks share their beloved native plant recipes, bringing color, culture, health, and life to your table.

61 | DESCRIPTION + PURCHASING OF INGREDIENTS

71	Acorn Bread
73	Acorn Dumplings with Venison Stew
81	Creamy Chia *Cacao* Pudding
83	Chia Cornbread
85	Chia Power Bars
87	Chia & Mesquite *Cacao* Energy Shake
91	*Cholla* Bud Succotash
95	Mesquite Tortillas
97	Mesquite & Chia Crispy Crackers
99	Mesquite & Chia Sage Pancakes
101	Native Superfood Cookie Trio: Mesquite, Chia, & Pine Nuts
107	*Nopales* Salad
109	*Nopales* Stir Fry
111	*Nopales* Tortillas
115	*Nopales* Tepary Bean Salad
117	Tepary Tart & Fruit Compote
121	Prickly Pear Cactus Frozen Treats
123	Prickly Pear Marinated Quail with Mesquite Rub
125	Rose Hip Jam
129	Stinging Nettle & Sunflower Seed Soup
131	Stinging Nettle Tea Medley
137	Marinated *Yucca* Blossoms
139	*Yucca* Petal Hash

149 | CONTRIBUTOR BIOS
RESOURCES & ACKNOWLEDGMENTS

NOTE TO THE READER: *Our book is intended as an informational and educational guide. The recipes, as well as the cooking techniques and preparation, are to inspire your own exploration and research into the wonderful world of native plants. Any recommendations for your health and well-being are for educational purposes.*

We are the

Founding Members

Craig Torres

Daniel McCarthy

Leslie Mouriquand

Barbara Drake

Our identity as a distinct people is derived from our place on Mother Earth. We need to educate a new generation that it's important to protect her. It's the only practice that will save human beings.

Native people managed the forest the same way as we do our gardens. It was just on a landscape scale. Pinyon pines, mesquite, oaks, nolina, yucca species, and so many of the other native plants were managed through harvesting.

There is something that draws me to the plants, both from a scientific/academic perspective, and from a spiritual-roots perspective. It's a glimpse of the lifeways of my ancestors. Goodness from the past for a hungry world.

Our ancestors worked so hard to preserve this land for us. Now it is our turn to teach the future generations the importance of saving and using native plants.

Chia Café Collective

Deborah Small

Lorene Sisquoc

Abe Sanchez

Native edible and medicinal plants offer us hope in an increasingly unpredictable climate. Not only are many of the plants incredibly nutritious, but they show remarkable heat and drought resistance, with the necessary resilience to adapt and survive.

We want to keep our traditions going, our earth and our people healthy. We're taking care of the land to create a healthy environment.

You are what you eat. Changes in life are hard, but decolonizing your diet and slowly replacing your food intake by a small percent with what your ancestors ate can make you feel better physically, mentally, and spiritually.

Cooking. Culture. Community.

Introduction by Craig Torres

FOR YEARS WE HAVE BEEN COOKING FOR THE COMMUNITY AND TEACHING CLASSES ABOUT OUR CONNECTION TO THE LAND AND TO NATIVE PLANTS. WE WANTED OUR ELDERS TO BE ABLE TO TASTE THEIR NATIVE FOODS, SO WE BEGAN GATHERING TRIPS TO GET FOODS AND MEDICINES AND PRESERVE THEM. WE ALSO STARTED A COMMUNITY PROJECT CALLED "PRESERVING OUR HERITAGE," AN INTERTRIBAL COLLABORATIVE WHICH HELPED US GATHER AND PRESERVE NATIVE PLANTS.

These projects have evolved into the **Chia Café Collective,** an organization whose strength comes from the knowledge and talents of its founding members: Barbara Drake, Lorene Sisquoc, Craig Torres, Abe Sanchez, Daniel McCarthy, Leslie Mouriquand, and Deborah Small. The Chia Café Collective continues to gather, process, and distribute plants to elders and others, but through our native food workshops and classes we now reach a broader community—working with various agencies, organizations, schools, and tribal communities.

We teach a "Traditional & Contemporary Native Plants Uses" workshop annually at Idyllwild Summer Arts, rotating each year between food, medicine, and utilitarian uses. And our aprons take us throughout California: giving workshops for *Cahuilla* Torres Martinez TANF tribal members, cooking for the *Haramokngna* American Indian Cultural Center, and fundraising for the **Dragonfly Gala** and Dorothy Ramon Cultural Center. Crowds jostle to try our mesquite pancakes (page 99) at the **Rancho Santa Ana Botanical Garden's Sage Festival.** People from all over southern California join the feast at the annual Malki Museum's **Agave Harvest & Roast**—and yes, we're there too!

We also give native plant food presentations and food cooking demonstrations for the Pimu (Catalina Island) Archaeology Field School every summer for U.S. and international university students. We transplant native plants in areas slated for development and cultivate them in our gardens in order to share seeds and cuttings with others. And when people grow the plants themselves, we provide information about propagation.

We are not food caterers, nor a formal nonprofit organization, but a grassroots group of individuals who share a philosophy that is action-based, a way of life connected to honoring all the indigenous people of southern California and our Mother Earth.

WHAT WE DO: ACTIVITIES, GOALS AND PHILOSOPHY

- Communicate the importance of cultural identity by reconnecting back to Mother Earth and developing reciprocal relationships with the natural environment and the "indigenous" of southern California

- Protect and restore California native plant communities and environments, while advocating for native plant landscapes on public and private spaces

- Reconnect California native plants as food, medicine, and utilitarian uses and the "gifts" they provide our human/non-human communities

- Provide information/education on plant resources for community harvesting and gathering of native plants

- Provide opportunities for mentoring/apprenticing youth leadership and fostering relationships between youth and elders. Provide plant foods for elders and tribal community members who do not have access or the ability to harvest and gather

Craig Torres *(Tongva)* sings this as a closing song to children for the *Tongva* program at Rancho Los Alamitos in Long Beach. His message to the children is, "even though we all come from different mothers, we all share ONE Mother Earth and we have the responsibility and obligation to take care of, honor, and protect her."

The *Going Home Song* is a song that was born at Puvuunga (the land on which California State University is built). It's a song about going home to our Ancestors—ancient ones who remind us of our connection to Mother Earth, the land that sustains us.

Going Home Song

from the Ti'at Society

Yamon hene, Yamon hene

Yamon hene, Yamon hene

Yamon hene, Yamon hene

Yamon hene, Hey yamon hene

Hey yamon hene nekiingaro

Yamon hene, Yamon hene

Yamon hene, Yamon hene

Preserving Our Heritage

THE BIRTH OF THE CHIA CAFÉ COLLECTIVE BEGAN WITH THE IDEA FOR A NATIVE FOOD BANK CALLED "PRESERVING OUR HERITAGE."

Barbara Drake: A few years ago, I was thinking about our elders. They didn't have a way to go out and gather their plants. Maybe physically they were unable to go, or didn't have transportation, or didn't know what areas they could go to.

I thought, *that's what our people need to do, our California people. On their journey through life, they need to taste things from their childhood, or from their mother's childhood, or their grandparents' childhood.* I talked to a friend who really liked the idea that we could organize ourselves to provide these food experiences for elders, and together we came up with the name: **Preserving Our Heritage.**

And these were our ideas...

We would gather in a traditional way. Our gathering trips would be a place to incorporate friendship, bonding, family values, making sure the children were with us, and preserving the plants for our elders so they could taste them again. We would freeze, can, and dehydrate them and then distribute them to the elders in a food bank. The more we all talked about it, the more we thought that a native food bank was something no one was doing, and that it would be a great idea.

Inspirations

Preserving Our Heritage had many inspirations, but a few remain dear to our hearts: Mrs. Levi *(Navajo)*, who always brought her beautiful little jars of hand-made foods to cultural events—plants like *cholla* buds and different types of cactus which she had gathered. And Linda Baguley *(Seneca-Cayuga)*, who told us about her people's path to the spirit world, which was lined with strawberries so that they could taste strawberries once more on their journey.

Honoring the Mother Earth Clan

THE CHIA CAFÉ COLLECTIVE WILL BE FOREVER GRATEFUL TO THE MOTHER EARTH CLAN MEMBERS: LORENE SISQUOC, CINDI ALVITRE, AND BARBARA DRAKE.

The Mother Earth Clan has given us many things over the years, including support, wisdom, and knowledge, and we try to live by the lessons they have entitled, "Clan Rules to Live By."

1. EVERYTHING IS SACRED AND ALL THINGS ARE ALIVE.
2. RESPECT YOUR ELDERS BY LISTENING AND LEARNING.
3. WHAT YOU DO WILL COME BACK TO YOU, GOOD OR BAD.
4. ALWAYS GIVE BEFORE YOU TAKE.
5. ONLY FOR SURVIVAL DO WE TAKE AN ANIMAL'S LIFE.
6. THE EARTH IS OUR MOTHER; DON'T HARM HER.
7. REMEMBER WHO YOU ARE AND WHO YOUR ANCESTORS ARE.
8. HONOR THE SACREDNESS OF YOUR MIND, BODY, AND SPIRIT AND DON'T POLLUTE THEM.
9. ALWAYS BE TRUTHFUL.
10. RESPECT OTHER'S PROPERTY.
11. BE A HARD WORKER (DON'T BE LAZY!)

Never Take
MORE THAN YOU NEED
by Craig Torres

My *Tongva* ancestors have inhabited this land for thousands of generations. In their creation stories, humans were the last to be made and were delegated the responsibility and obligation to maintain a sacred balance with nature and all life on Mother Earth. A reciprocal relationship was born, and our people enjoyed the food, shelter, clothing, and medicines that were provided by Mother Earth. The landscape was saturated with the dance of life; ceremonies, laughter, traditions, people raising families, songs, and the seasonal gathering of plants. Trade networks criss-crossed these landscapes, connecting communities during times when resources were scarce.

Nature was not viewed as "natural resources," endlessly being used up. This was the way of life, the laws of nature, but this philosophy would find itself under attack by the first wave of foreigners who permanently colonized in the late 1700s. Under Junipero Serra, the Mission System was created in California, and its policies reduced hundreds of Native communities into controlled, centralized locations. This irreparably severed the ceremonial, inter-marriage, and trade networks that had enabled Native communities to survive on the vast bounty of the land for countless generations.

As permanent European colonization began, a severe disruption of the Native people, animals, plants, and the land itself followed. In the Los Angeles basin, the landscape was altered by cattle and livestock grazing under the Spanish government's *rancho* land grants. Many of the native grasses and plants (such as chia) were drastically eradicated from the vast prairie valleys and plains where they once dominated.

Spanish policies banned the traditional, seasonal burning practices of landscapes by Native peoples. As a result, cattle and livestock had more grazing plants, but Native people and animals were left with a crippled seasonal yield and harvesting cycle. These practices, along with small-scale agriculture, would change the landscape forever.

Early European colonizers failed to recognize the value of Native landscape management practices that had shaped and conserved the environment for hundreds of generations. This mindset was carried through the Spanish occupation, the American occupation (beginning in 1848), and has continued into the 21st century.

In his article, *"Century of Destruction: Environmental Devastation of the San Gabriel Mountains"* on the KCET website, author Daniel Medina writes, "The nineteenth century was a brutal era of exploitation of the San Gabriels. Its valuable natural resources were at the mercy of an expanding populace that believed the wealth of nature to be inexhaustible. Abuse of its timberland, watershed, and wildlife would fray its balanced ecosystem. Before the Forest Reserve Act of 1891 established them as federally protected lands, the public domain of the San Gabriels was subject to unchecked resource demands and careless destruction." Unfortunately, this was not relegated to just the San Gabriel Mountains but to the entire Los Angeles basin.

The ocean landscape would also suffer severe losses and near extinctions. Sea otters off the Los Angeles coast and southern Channel Islands were hunted to near extinction in the early 1800s by American and Russian poachers and smugglers. The Mexican government later outlawed sea otter hunting along the California coasts, but enforcement on the relatively remote Channel Islands was undoubtedly difficult and rather ineffective. The trade economy produced by the selling of otter hides to the Asian market and their demand for a high price was worth the risk. They continued to over-hunt well into the American period. By the 1850s, sea otters had nearly been eradicated from California, and as a result, dramatically effected shellfish populations and kelp forest habitat, disturbing the natural ecosystem of the ocean they once inhabited.

In 1848, the discovery of gold brought immigrants into northern California by the thousands, seeking fortune and wealth from the land. In southern California, specifically the Los Angeles basin, it was "Black Gold" that was extracted as a commodity from Mother Earth. In one oral story, a *Tongva* community in the San Pedro/Wilmington area near the Los Angeles harbor was massacred for refusing to move from their land—land which sat upon abundant oil reserves. In the 1890s Los Angeles experienced its first oil boom, which led to the 1920s oil spills that damaged hundreds of acres across the Los Angeles basin.

In the latter part of the 19th century, logging and timber extraction, particularly in the San Gabriel Mountains and foothills, were exploited by settlers who profited from harvesting timber and saw-pit operations. It wasn't until the Forest Reserve Act of 1891 established them as federally protected lands that they were protected from careless overexploitation.

The attack on the landscape in southern California continues: the extinction of the grizzly bear, DDT spilling in the Los Angeles harbor, near extinction of abalone (twice), urban sprawl vs. mountain lions and coyotes, rampant agricultural farming, and the destruction and alteration of wetlands and watersheds.

AS AN ELDER ONCE REMINDED ME MANY YEARS AGO...
"WHEN WE IGNORE NATURE, IT WILL GO AWAY." ONE OF THE WAYS WE CAN
PREVENT THIS IS BY HONORING THE NATIVE PLANTS OF THE LAND.

Make a Change

"Never take more than you need to survive on Mother Earth," and "Ask permission before taking," and "When you take something . . . give back." Some of these actions take the form of dispersing seeds when harvesting, so that the plants have a chance to reproduce and endure. And when you're gathering plants, always leave some behind to feed others who come after.

Native Foods

AND WHY THEY MATTER

by Deborah Small

"When a person eats acorns, mesquite pods, tepary beans, or prickly pear cactus, the special dietary fiber in these foods slows down the release of sugars into the bloodstream and extends the period over which nutrients are absorbed into the body, while increasing insulin sensitivity. . . . The recovery of the people is tied to the recovery of food, since food itself is medicine: not only for the body, but for the soul, for the spiritual connection to history, ancestors, and the land."

—Winona LaDuke, *Recovering the Sacred*

We are the **Chia Café Collective,** a grassroots group of southern California tribal members and collaborators committed to the revitalization of Native foods, medicines, culture, and community. Our work honors the vast traditional knowledge and spiritual relationship to the land, and the nutritive and medicinal bounty the land offers us. Through workshops, classes, demonstrations, and native foods celebrations, we focus on ways to re-incorporate Native food plants into our daily diets to take back responsibility for our health and well-being. We hope our recipes can help you reconnect with the land through gathering, gardening, and cooking Native foods. We promote an ethic of gathering and cultivating native plants in a manner that is sustainable, and we stress the importance of preserving native plants, plant communities, habitats, and the land for the future generations of all species.

Ancestral food plants offer us hope in an increasingly uncertain climate. Not only are they incredibly nutritious, but they also show remarkable heat and drought resistance, with the necessary resilience to adapt and survive in the face of the unpredictability of climate change. According to ethnobiologist Gary Nabhan, "prickly pears are among the most water-use-efficient crop plants in the world." Once established, they require no supplemental irrigation. They're incredibly easy to grow, with little or no maintenance.

Tepary beans "are among the most drought-adapted annual legumes on earth." The crop is ready to harvest in just 55-65 days. And mesquite has an extremely deep taproot system and expansive lateral roots for survival in arid regions.

These native plants also matter because they help us take back responsibility for our health and well-being. Drought resistant *cholla* cactus flower buds pack a nutrient punch, especially their highly absorbable calcium, a boon for lactose-intolerant people. Mesquite and acorn are two of the most effective foods tested for controlling blood sugar levels and diabetes. Chia seeds are high in protein and fiber, and their mucilage is of great assistance for our digestive system. Cleveland sage has one of the most intoxicating fragrances of all the southern California native plants. Used in potpourri and sachets, it's also an aromatic and tasty addition to mesquite pancakes or as the primary "green" for pesto.

We encourage everyone to spend quality face-time with the plants in their natural habitats, and to make friends with the sage scrub and chaparral, the oak woodlands and riparian streams, the juniper pinyon forests, and the deserts. We also encourage you to venture into the kitchen with new ingredients in hand, to concoct something delicious and healthful, and through these foods to reconnect with your ancestors, family and friends, and the earth.

IDYLLWILD ~ p 24

CATALINA ISLAND ~ p 37

TORRES MARTINEZ ~ p 42

MALKI MUSEUM ~ p 49

SHERMAN INDIAN HIGH SCHOOL ~ p 54

"The Cahuilla people came here to Idyllwild to gather their acorns, and their spirit is all over these lands. In the meadow, there is a beautiful outcropping of grinding rocks. If you're really quiet, you can hear the Cahuilla ladies talking and laughing."

IDYLLWILD

Annually, during the last weekend in June, members of the **Chia Café Collective** give a California Native Plants Workshop for the Idyllwild Summer Arts Program. A spectacularly beautiful setting, this is the ancestral land of the *Mountain Cahuilla* people. We introduce students to the vast edible and medicinal bounty that surrounds us in southern California. And we share what we're really passionate about—the incredibly diverse properties of California plants that are used as food, medicine, fiber, dye, and shelter.

On the first day of the workshop, students and instructors meet in a meadow next to grinding rocks that have been used for thousands of years. As introductions are made, it's obvious that we are all friends for whom sharing is a way of life. Our respect for each other is palpable, and we speak of each other as mentors as well as collaborators. After introductions, Barbara, Lorene, and Craig smudge everyone with white sage and thank the Creator. Barbara explains that the smoke helps take their prayers upward. We ask the students to join us by singing three songs: the *Willow Song* for the plants, the *Ancestor Song* for those who came before us, and the *Ocean Song* for the waters. Elderberry clapper-sticks keep the beat.

During the workshop, students participate in the entire process of food preparation. They gather rose petals for *yucca* blossom salad topped with pine nuts. They use gloves and tongs to harvest stinging nettle leaves for vegetable dishes and soups. They mix *masa* with stinging nettle, amaranth, and *nopales* and then roll out fresh tortillas, which they cook on a *comal* on an outdoor camping stove. Acorn bread and mesquite cookies are baked as well. Everything is prepared with great care, enthusiasm, and passion.

Workshop topics include:

- Embracing the use of modern tools like juicers, blenders, food processors, electric grinders, dehydrators, and freezers. The ultimate goal is for Native people to once again incorporate healthy foods into their daily diets.

- Promoting cuisine with Native foods as the primary ingredient, but joined with other foods for added nutritional value.

- Inspiring social justice for communities denied their traditional foods and cultural practices.

- Teaching the protocol and etiquette of gathering.

- Encouraging students to cultivate native plants in their gardens or in containers on decks, porches, and windowsills, and to become active in community gardens.

We designed the Idyllwild workshop as a stepping-stone to help people sustain healthy and creative lives. By empowering them through their food and medicine choices, they cultivate a deeper relationship to the plants, animals, and earth.

We want our students to leave the workshop with a better understanding of the botanical, ecological, and cultural aspects of our complex world.

Craig Torres

"When we harvest plants there's a certain protocol and etiquette to follow. We leave offerings or we pray to the plants. We ask them for their medicine and for their healing before we take them, because that's a big part of our connection to the land. Over time, you begin to develop a relationship with the plants that you wouldn't otherwise have."

Lorene Sisquoc

"These are very important plants to us. They provide not only our nutrition but they can also save some of our people's lives; most of these foods help with diabetes and the other negative things that come along with the way we eat today."

Barbara Drake

"In the past, we lived within our environment, respecting it. We knew what it would give to us and we knew our role: To take no more than we could use. To share what we gathered with everyone. We all knew that if we didn't do that, it was endangering the life of everyone."

QWINYILY :: BLACK OAK

WI'ASILY :: LIVE OAK

PAWISH :: SCRUB OAK

WI'AT :: CANYON OAK

Katherine Siva Saubel, a Cahuilla elder who had been raised since childhood with Cahuilla traditions, was our friend and guide. The last five years of her life, she travelled up here to Idyllwild and sat in the meadow with us. She brought her own plants and talked about them, and it was the greatest honor to have a person like her here with us. When I'm here, I feel Katherine's presence, and I remember how she taught us about the plants and the words for them in the Cahuilla language.

> When we come to Idyllwild the ancestors watch us. And they're grateful that we care and that we come. I put an offering of tobacco on the rocks to thank those ancestral *Cahuilla ladies*, because they're still here. So when you're walking on this beautiful land, think of the *Cahuilla*.

Barbara Drake: I am a *Tongva* elder and culture keeper for our tribe from the Los Angeles basin. Our ancestors built the San Gabriel Mission for Spain. And of course we lived on that land for thousands of years before that. I've had wonderful people surrounding me my entire life. I owe all this to my mom, who as children took us out gathering. She fed us off the land. And she doctored us. I didn't go to medical doctors until I was 16 years old. My appreciation and love of these plants is from my mother.

The *Cahuilla* people came here to Idyllwild to gather their acorns, and their spirit is all over these lands. In the meadow, there is a beautiful outcropping of grinding rocks. If you're really quiet, you can hear the *Cahuilla* ladies talking and laughing. It was a tedious task to grind those thousands of acorns, but they did it with great happiness. They would sing their songs and tell stories. Processing the acorns for the people they loved made the work go much easier.

LESSONS FROM IDYLLWILD

MAKING ELDERBERRY MEDICINE TUBES

Barbara Drake *(Tongva):* For our medicine containers, we use 5-inch lengths of elderberry tree branches. It's the same wood that we use to make our clapper-sticks and our flutes. After you peel off the bark, you hollow it out. It's got a soft center. You use a skewer to go all the way through to remove the pith, because later you'll plug up the bottom of the hole with wax. After that, you sand the tube as smooth as can be. Then you're ready to decorate it with a wood burner. When you're finished, you put your medicine in there.

I grow the fiber plant (dogbane) in my garden, and I bring enough for everyone to make cordage to hang their elderberry medicine containers around their necks. Abe Sanchez brings natural pigments so students can stain their elderberry tubes with red and yellow ochre.

ELDERBERRIES IN MY PRAYER POUCH: THE SISQUOC WAY

Lorene Sisquoc *(Cahuilla/Apache):* I use elderberries in my prayer pouch. I like to have a mixture of my prayer plants and I put dry elderberries as the "food part" to give out of my offering pouch. I use a handful of this mix when I gather a plant or when I'm praying.

I also use sage and tobacco and cedar. Or sometimes just sage and tobacco. And then some elderberry. Some people say you don't mix it all. Some do. So that's my way. It's not from any one of my tribes. It's just my way. It's the Sisquoc Way.

Rita Esmeralda Naranjo proudly wears the elderberry medicine tube she made in one of our workshops.

LESSONS FROM IDYLLWILD

MAKING "WONDER SALVE"

Craig Torres *(Tongva):* Some of my most cherished childhood memories are the times we spent on my uncle and aunt's Acosta Ranch near Lake Elsinore. The small family ranch house was sprawled over a few acres of open land that seemed to stretch for miles. Aside from the food and drink and typical family ranch socializing, I remember running all over the land, playing and exploring with my cousins. I also remember the smell of native plants that permeated the air— California white sage, California bay, California sagebrush as well as the smell of the little stream down the way, edged with tules and cattails where we sometimes caught frogs and polliwogs. I never dreamed that as an adult, some of these smells would come together into a salve that I could use to heal skin conditions. It's a salve I trust more than any store-bought brands. Besides, the smells that emanate from my salve carry those healing childhood memories of Mother Earth.

Many of the chemical constituents of these plants are anti-infective, anti-microbial, anti-inflammatory, and analgesic, so it is really more than just "anti-fungal." Applied topically, the salve is excellent to remedy skin conditions caused by bacterial *staphylococcus* infections that are visible on the skin and have a reddish, swollen, and tender area at the site of infection. It's good for fungal infections such as athlete's foot, jock itch, ringworm, yeast infections, and candida-induced "sweat rash" or "prickly heat." And it's amazingly helpful in combating eczema, herpes, cold sores, psoriasis and contact dermatitis.

Wonder Salve

by Craig Torres

INGREDIENTS

- 1 C   *Creosote (Larrea tridentata)*
- 1 C   *California white sage (Salvia apiana)*
- ½ C   *Yerba Mansa (Anemopsis californica)*
- ½ C   *California bay (Umbellularia californica)*
- 👁   *olive oil*
- 👁   *beeswax (pellets)*

All of these plants can be purchased in dry-leaf form online or at herbal stores.

INFUSING OIL WITH THE MEDICINAL PLANTS

Crush or chop dry plants into a "rough chop" and mix together. Place in a quart jar, leaving at least a 1-inch space at the top, and cover plant material with olive oil. Cap the jar, place in a brown paper bag, and leave outside in the sun for 1 week or longer. Shake the jar every other day. After plants have infused into the oil, strain plant material using muslin cloth. Plant material can go into the compost pile.

MAKING THE SALVE

Place the medicinally-infused oil in a small pot on low heat, but do not boil. Add a small amount of beeswax (pellets work well) and stir until melted and infused with the medicinal oil. Test the consistency by dropping a teaspoon of the medicinal oil/wax liquid into cold water; it should feel like chilled butter. Add more wax if the salve is too thin or more oil if the salve is too thick.

PACKAGING & STORAGE

Pour into lidded containers and let sit until the mixture solidifies. Store out of the sun/heat.

By the end of the workshop, new friendships blossom and we exchange hugs and promises to meet again.

Barbara Drake and Petee Ramirez pictured here.

CATALINA ISLAND

Every year, several members of the **Chia Café Collective** travel to Catalina Island to give native plant food workshops and cooking demos for the *Pimu* (Catalina Island) Archaeology Field School. We arrive Friday, eat tacos and sip margaritas in Avalon with the Field School teachers and staff, then give a slideshow presentation on California native plants and ethical gathering protocol and practices. At night, we stay in the camp with platform tents under skies teeming with stars. Saturday is our all-day cooking workshop in the outdoor kitchen looking out at the Pacific Ocean that surrounds us. On Sunday, we hike along the coast to gather seagrass, share snacks and stories, and swim.

Pimu

ARCHAEOLOGICAL FIELD SCHOOL

NATIVE FOODS MENU

MEAT
Marinated Quail in prickly pear juice, rubbed with mesquite
Shellfish Stew with seaweed, welk, and other shellfish

SOUPS
Venison & Coreopsis Soup
Seafood Soup

SIDES
Amaranth Green Tacos
Stinging Nettles Dip with White Tepary Beans
Seaweed Salad
Cactus & Tepary Beans
Acorn Soup

BREADS
Mesquite Tortillas made from mesquite flour, wheat flour, butter, and baking soda

DRINKS
Cactus Fruit Drink
White Sage Lemonade
Manzanita Berry and Lemonade Berry Drinks
Chia Seed Agua Fresca

DESSERTS
Brownie Bites (no-bake and gluten-free) made with chia seeds, date paste, black walnuts, *yerba buena,* and *cacao* paste

White Tepary Bean Pie topped with elderberry sauce inside a mesquite/almond meal/chia crust

(Add stars, laughter, and the sounds of the ocean.)

Torres Martinez Native Foods Workshop

In April 2012, we had the privilege of working with the *Torres Martinez Desert Cahuilla* at a weekend Native Foods Workshop at their cultural center in Hemet, California. The workshop was organized by Daniel Salgado, who had attended the workshop, "Native Plants for Food & Utilitarian Purposes," taught by us at Idyllwild Arts Center the preceding summer.

The *Torres Martinez* workshop, a celebration of southern California Native foods, promoted what we like to call a Native-strong cuisine: native plants used as the primary ingredient, but joined with other foods for added nutritional value, flavor, and beauty.

Day 1: Food Prep

TO BEGIN, *TONGVA* ELDER BARBARA DRAKE OFFERS A VERY SPECIAL THANKS TO THE *CAHUILLA* PEOPLE PRESENT. "WHEN THE *TONGVA* RAN AWAY FROM THE MISSION, THE *CAHUILLA* WOULD TAKE THEM IN. THE SPANISH SOLDIERS WOULD ONLY PURSUE THE *TONGVA* INLAND FOR 40 MILES."

Daniel McCarthy then introduces the students to *Temalpakh,* Katherine Siva Saubel and Lowell Bean's collaborative ethnobotany of *Cahuilla* people. "It's a food Bible, and not just for *Cahuilla* people. Many Natives used these or similar plants."

Barbara Drake brings roasted pinyon pine cones from last year's pinyon pine cone gathering to show the workshop participants. "If a mother died in childbirth," she tells the students, "they would grind the pine nuts and feed the baby pinenut milk. It is that nutritious and easy to digest."

Chia is one of Craig Torres' specialties. To make pomegranate chia gel, he has the students soak the chia seeds in pomegranate juice. The chia absorbs the juice, the mixture thickens, and then the students spoon the chia gel into small paper serving cups. "It's really good," Stella Rico tells us. "You can really taste the pomegranate."

Abe Sanchez's insect specialty is grasshopper tacos. First, the students roast the grasshoppers, adding onions, garlic, and tomatoes and serve on blue and yellow corn tortillas. For this occasion, Abe also adds sautéed moth larvae to his insect repertoire. Historically, *Cahuilla* ate the white-lined sphinx moth, *Hyles lineata,* which pollinates the seductive, sacred datura, *Datura wrightii,* a flower that opens at twilight and blooms for a single night. The empathetic Haydee Rico is not so sure about eating insects. "I feel bad for the larva," she tells us.

Abe also teaches the students how to prepare rabbit for grilling. "We didn't capture these particular rabbits using traditional rabbit sticks," Abe tells everyone, "so we aren't able to skin the rabbits for the fur to make blankets or capes like your ancestors did." The students love the rabbit. Johnny Guzman and Robert Rico both agree. "Rabbit is WAY better than chicken!" Johnny adds, "It's delicious and nutritious. It's even better than cow."

Barbara has students help her set up one of her famous tea stations, where everyone creates their own teabags with rose hip petals (for Vitamin C), pineapple weed, and horsetail (for kidney and bladder), yerba santa (for respiratory challenges), and elder flowers (for colds and flu).

Deborah Small brings 12 pint jars of prickly pear juice that she froze last October for workshops such as this one. The students mix organic limes and lemons with the prickly pear juice, throw in a few handfuls of chia, then add water and ice to make a delicious prickly pear lemonade. The lemonade is a big hit. Over the week-end they made three 5-gallon batches.

Day 2: Feast Day

ON SUNDAY, THE STUDENTS INVITE THEIR FAMILIES TO THE NATIVE FOODS FEAST. AFTER A BLESSING, CRAIG INTRODUCES *THE ANCESTOR SONG*. "YOU ARE THE NEXT GENERATION. IT'S UP TO YOU TO CARRY ON OUR TRADITIONS, TO HONOR OUR ANCESTORS. FOR US, FOR ALL INDIGENOUS PEOPLE, IT'S IMPORTANT TO SING TO OUR ANCESTORS, TO THOSE WHO CAME BEFORE US. THIS SONG ASKS THE ANCESTORS TO ACKNOWLEDGE US, TO ACKNOWLEDGE OUR HEARTS, OUR SPIRIT. TO ASK THE ANCESTORS TO BE WITH US."

After singing the *Ancestor Song*, Craig introduces *Going Home*. "It's about going back to your roots, never losing touch with where you came from, with your ancestors. As elders, we want to give our knowledge away. You are the generation that will take over from us. We always hear that the young people are not interested. They don't want to learn the plants. They don't want to learn the songs. But here at Torres Martinez, that's not the case."

Barbara orchestrates a beautiful moment as the students line up behind the long tables to serve their elders, parents, younger siblings, friends, and teachers. They serve everyone generous portions of the venison soup, nettles, grilled quail, acorn bread, chia cornbread, *yucca* salad, purslane salad, and chia power bars that they have helped to prepare.

And that's not all. On a side table, people help themselves to *Cahuilla* acorn soup (*wiiwish*), as well as agave hearts roasted at a pit at the nearby *Cahuilla* Malki Museum for their annual Agave Food Festival.

The young students participate in the entire process of food preparation.

There are plenty of leftovers for everyone to take home to share with their families. We all want to savor the vegetable venison stew, the acorn bread, and the chia power bars in order to extend the week-end food festival a bit longer.

MALKI MUSEUM

Each spring, on the Morongo Indian Reservation near Banning, the Malki Museum holds an annual Agave Harvest & Roast. For many years, Daniel McCarthy, various *Cahuilla* elders, and the Malki museum have been sharing knowledge about agave as an ancient food and fiber for people of this region. At this annual event, the Malki offers an opportunity for the general public to taste agave and for people to come together to celebrate this plant and other native plants of vital importance to the southern California Native communities. The Malki was founded in 1964 by Jane Pablo Penn and Katherine Siva Saubel, *Cahuilla* scholar, historian, and revered tribal elder.

We are proud that two of our **Chia Café Collective** members, Lorene Sisquoc and Daniel McCarthy, currently sit on the Malki Museum Board of Directors. And we are thrilled to be involved with the harvest and roast: the first week-end, gathering the agave hearts, and the second week-end, demonstrating and volunteering in the kitchen for the Roast. This event allows us both to share what we know and to learn!

SHERMAN INDIAN HIGH SCHOOL

Lorene Sisquoc *(Cahuilla/Apache)*, one of our founding Chia Café Collective members, has a unique opportunity as a Cultural Educator at Sherman Indian High School in Riverside, California to teach her Native students about making traditional foods part of their lives again.

Lorene: We're growing native plants here at Sherman Indian High School to bring back Native foods. It's an important part of our cultural programs. We designated a Native foods tasting day during Native Heritage Month, and made it an annual event. I work with students from 70 to 100 different tribes here at the school. We want the student body to take pride in each of their culinary specialties.

I asked the students to think of their favorite traditional dish. And everybody replied: frybread. And I said, "besides frybread." Pretty soon, one student said they liked blue corn meal, and another, blue corn mush, or piki bread, or buffalo stew, or wild turnips. One little boy got so excited that he said he was going to call his grandma in Alaska and have her send some food.

"My mouth just waters thinking about it," he told us. Weeks later he comes into class and says, "Ms. Sisquoc, I got my box from my grandma. It came in."

And it's dried seal blubber and it didn't look that appealing, but he was just so excited. And the other students got to taste it and it was exciting. It was fun. He brought in a little bit of home to the class and shared that with everyone.

We encourage our local community to bring in acorn mush and chia from this area to share. The students from tribes outside California bring in their foods too, and some days it turn into a nice, big feast.

In September of 2008, Chef Nephi Craig *(Apache)* came to visit for California Indian Day. I had heard him speak at the Heard Museum the year before and thought *wow,* this is exactly what what we're trying to do. In his presentation he talked about the sacredness of food and how our people use their own culinary ways of cooking. He came to Sherman for four days and worked with our culinary teacher to give lessons. On California Indian Day everyone did tastings, the students learned about plating, and they all helped. It was a dream come true, and I'll never forget it.

We have a traditional plant-use class here at Sherman, where I incorporate Native food-use and planting, as well as integrating those lessons into my Basketry class and my Native Traditions class. I feel it's so important that we have this here because of the history of boarding schools—a history of taking our foods away and bringing the kids here in the early days to become farmers of non-traditional crops. [They told them to] forget about their traditional plants and traditional gathering. Forget about those acorns, pine nuts, and mesquite. Instead, garden and milk cows.

So many of our culinary secrets and specialties were not passed down because the kids weren't home to learn them. Luckily, a few were still home, or some just made a point to learn anyway and kept the knowledge of native plant foods. The boarding schools had a big negative impact on every aspect of our culture, from the baskets, to the foods and medicines, and on and on. Which is why it's so important to teach that native culture is significant and important. Language, basketry, plants. All those things need to be taught here; shared, shown, and highlighted. That's why I like to do classes and courses and presentations here at Sherman, because I get to help that knowlege come back.

Students might not think that they know things, but then they say, "Oh, my grandma does this." They'll remember that thing and do it. Or, they'll get the idea to find out, and they'll ask, "Well what did we do? What did we use? What's one of our good foods?"

Recipes

Featured

NATIVE FOODS: DESCRIPTION & PURCHASING ~ p 61

THE STORY OF ACORN ~ p 65

GATHERING CHIA ~ p 75

YUCCA MOTH DANCER ~ p 133

I FOUGHT TYPE-2 DIABETES AND WON ~ p 142

WELCOME TO OUR KITCHEN

FOR MOST PEOPLE, COOKING IS AN ADVENTURE, BUT WITH OUR NATIVE FOOD RECIPES, GETTING YOUR INGREDIENTS CAN BE EQUALLY CHALLENGING. WORTH IT? WHEN YOU BITE INTO YOUR FIRST ACORN DUMPLING, YOU'LL HAVE YOUR ANSWER.

Our Native recipes use ingredients which can be challenging to get because they come from plants and animals which are specific to the southwestern United States and northern Mexico. Luckily, we're here to guide you on your Native food adventures with tips on purchasing, descriptions of unfamiliar ingredients, and detailed processing of special ingredients.

Our recipes reflect the tastes and culture of the Native people here, and perhaps most importantly, the seasonal rhythms that Native people live by; you can only get fresh *cholla* buds in the spring, because that's when they're ready to harvest.

We will introduce you to a new language of ingredients, an exotic blend of Spanish and Native words. Words like "*cholla bud, nopales,* mesquite, and *wiiwish*" will make their way into your kitchen and fill your pots with healthy, colorful food.

COOKING ISN'T JUST ABOUT EXACT MEASUREMENTS; IT'S FROM THE HEART. OUR COOKS DO A LOT OF EYEBALLING INGREDIENTS, SO YOU'LL SEE THAT INDICATED THROUGHOUT OUR RECIPES.

———— ICONS ————

Ingredients that need special processing before using.

Decription and purchasing of special ingredients.

—— ICON ——   —— ICON ——

Description D+P Purchasing

WE LIKE TO GET OUR INGREDIENTS FROM COMPANIES OR INDIVIDUALS THAT HARVEST THEIR PLANTS AND ANIMALS IN A SOCIALLY AND ENVIRONMENTALLY CONSCIOUS WAY. WE ENCOURAGE YOU TO DO THE SAME.

ACORN FLOUR	Nuts from oak trees. In our recipes, it's used in flour form, which involves special processing before it can be used.	*Slightly sweet with a mild, nutty taste.*
CACAO POWDER	The fruit of *cacao* trees contain small beans. *Cacao* powder is ground from the beans.	*Richly flavored, bitter chocolate taste.*
PRICKLY PEAR CACTUS	The small new-growth of cactus pads *(nopales)* or fruit *(tunas)*.	*New growth pads have a slightly tart flavor with a texture similar to okra. Prickly pear fruits are sweet, refreshing, and mildly tart, and come in magenta, yellow, orange or green colors.*
CHIA	Seeds from the chia plant. Can be eaten raw, whole, ground, or roasted.	*Raw seeds have a slight nutty flavor but the taste is enhanced if roasted in a skillet over a low flame.*
CHOLLA BUD	Flower buds from the *cholla* cactus.	*Cooked buds taste like a cross between artichokes and asparagus.*
MASA HARINA FLOUR	Field corn that is dried, treated with slaked lime, and then ground into flour.	*Lends a savory, nutty flavor to freshly made tortillas.*
MESQUITE	Seed pods from the honey mesquite tree and other species. In our recipes, we use mesquite flour.	*Sweet, earthy, molasses-type flavor.*

Companies:
Sue's Acorn Café and Mill (black oak acorn) **www.buyacornflour.com**
Oaklore Acorn Flour (northern red oak acorn) **www.oakloreproducts.com**
Akoshiilaa Sheila! Navajo Gallery (emory, gambel, and live oak oak acorn)
akoshiilaa.sheila@hotmail.com, (928) 814-8847 or (928) 814-8391

Company: Sunfoods **www.sunfood.com**
Also widely available online and at health food stores.

Companies:
Arizona Cactus Ranch **www.arizonacactusranch.com**
Desert Tortoise Botanicals **www.desertortoisebotanicals.com**
Pads and fruit are also available in most Mexican/Hispanic markets.

Widely available at Costco, health food stores, or Mexican/Hispanic markets.

Companies:
San Xavier Co-op Farm **www.sanxaviercoop.org/ciolim.html**
Tohono O'Oodham Community Action (TOCA) **(520) 383-4966** or **www.tocaonline.org**
Native Seeds/SEARCH **www.nativeseeds.org**
Flor de Mayo Arts **www.flordemayoarts.com** (available seasonally)

Widely available at Mexican/Hispanic markets and the ethnic sections of other markets.

Companies:
San Xavier Co-op Farm **www.sanxaviercoop.org/ciolim.html**
Native Seeds/SEARCH **www.nativeseeds.org**
Sunfoods **www.sunfood.com**

Description D+P Purchasing

	Description	Purchasing
QUAIL	A small, ground-dwelling bird with a distinct feather plume on its head.	*Tender, delicate-tasting meat.*
ROSE HIP POWDER	Small fruits from the rose bush which have been ground into a fine powder.	*Very tart, with a sweet aftertaste.*
CLEVELAND SAGE	Leaves of the Cleveland sage plant.	*Savory, with a slightly bitter undertone.*
STINGING NETTLE	New spring growth or tops of young nettle plants.	*Tastes like spinach when cooked, only more delicious. Quantity greatly reduces when steamed or cooked.*
TEPARY BEANS	Sonoran Desert domesticated bean.	*Nutty bean flavor; white beans taste sweet, brown more earthy.*
VENISON	Meat from a deer.	*Gamey, very lean meat flavor similar to beef.*
YUCCA BUDS / BLOSSOMS	Buds and blossom from yucca stalk.	*Flowers taste like asparagus when cooked. Buds are tart.*

Often available at Asian markets and specialty butcher shops. Or befriend a hunter who may want to share their hunt.

Company: Mountain Rose Herbs **www.mountainroseherbs.com**

Cultivated varieties widely available at California native plant nurseries.

Company: Mountain Rose Herbs **www.mountainroseherbs.com**
Or gather wild in gardens where it frequently volunteers.

Companies:
San Xavier Co-op Farm **www.sanxaviercoop.org/ciolim.html**
Native Seeds/SEARCH **www.nativeseeds.org**
Sunfoods **www.sunfood.com**

Often available at Asian markets and specialty butcher shops. Or befriend a hunter who may want to share their hunt.

In most states, a permit is needed for wild collection of *yucca,* which is why we don't know of any commercial sources for it. We gather the blossoms on the land of friends. To cultivate your own, plants are available at many California native plant nurseries.

THE STORY OF
Acorn

by Tima Lotah Link

Throughout human history, across almost every continent, acorns have provided food for people—it's a food that unites us all.

In the past, California Natives prepared acorn as breads, biscuits, patties, dumplings, coffee, cheese, and nuts. But the most common way to eat acorn was to cook it with water to achieve whatever thickness that a particular cultural area preferred: thin like soup, thicker like pudding, or firm like Jello. Similar to rice, noodles, or potatoes, this food was eaten as the central part of a meal, its delicate flavor enhanced by meats, vegetables, nuts, and fruits.

Early anthropologists exposed their cultural prejudices by labelling this native food as "acorn mush" or "acorn gruel"—unflattering words that painted a picture of a watery, tasteless food. In reality, names like "acorn bisque" or "acorn soup" better capture the nutty, delicate flavor of acorn. Native people across California simply use the word in their language for this exceptional food; we have shown only a few of the more than 187 different ways to say "acorn soup" in California.

Today, while Native people still eat acorn in its traditional forms, new and exciting ways to prepare acorn are being born, and people everywhere are re-embracing the flavor, culture, and nutritional value of this remarkable food.

SPECIAL PREPARATION

ACORN FLOUR

Today you can purchase acorn flour made from oaks growing around the world. We have our favorite suppliers of course, which are listed in the Description & Purchasing section (page 61). But for those of you who gather your acorns, it's worth noting that there are more than 13 species of California oaks which can be used for acorn food. Each species produces acorns which have different sizes, storage potential, ease of shelling and pounding, degree of bitterness and leaching time, texture, taste, and nutrition.

In a dizzying display of culinary experience, Melba Beecher *(Mono)* explains how her ancestors relied on six different kinds of oaks for acorns: "Mostly they used black oak and Spanish oak [together]. Water oak is used for emergency because it doesn't thicken up. Live oak is used the same as water oak. It's too hard to shell. Scrub oak is used too but that and Spanish oak don't keep a long time. Blue oak is mixed with black oak. You have to keep pounding it. So it is not used as much as black oak. If you only used Spanish oak [the soup is] almost white when you cook it."
(M. Kat Anderson, 2009: *Historical Native American Use, Harvesting, and Management of California Oak Communities*)

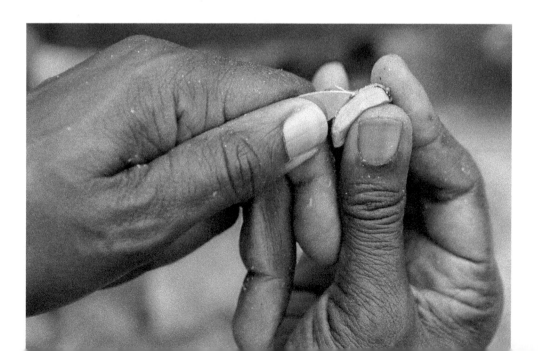

"The Acorn Lady"

LOIS CONNOR BOHNA (MONO)
Master acorn-maker and basketweaver

"Oak trees are everything to us and we honor them by singing prayers when we gather their acorns. When acorns are made into food, 90% of that should come from a spiritual place inside you."

ACORNS CONTAIN LARGE AMOUNTS OF PROTEIN, CARBS, AND FATS, AS WELL AS CALCIUM, PHOSPHORUS, POTASSIUM, AND NIACIN.

LEACHING = removing the natural tannic acid from the acorn flour with water. Tannic acid isn't harmful, but it's extremely bitter, which is hard on your stomach and your taste buds.

When you buy acorn flour from a supplier, you should request that it come unleached. This will allow YOU to leach the flour, which guarantees that it will *(a)* have a smooth, bitter-free taste and *(b)* retain all of its nutritional content (stirring or overhandling while leaching can deplete the flour of fatty omegas).

If you cannot get unleached flour, you should still check to see that the supplier has leached it correctly to remove all tannic acid. Simply wet your finger, dip it in the flour, and taste. Bitter? You have to leach it more. Not bitter? You're ready to use it in your recipe as is.

Your acorn flour should be fine-grain, like baking flour, but if it's not, run the dry flour through an electric grinder for a finer grain.

ACORN FLOUR: PREPARING TO LEACH

There are many ways of leaching acorn, but the following yields the quickest and best results:

1. Using a nail, punch **lots** of holes in a large, flat baking sheet (the holes should all be punched downwards, creating tiny funnel-shaped holes.) You're making yourself a colander that will allow water to strain easily through it and most importantly—it's FLAT. A flat surface will allow you to spread the acorn flour out in an even thickness. When water is poured over this evenly spread flour, it will drain through the flour consistently, and cut your leaching time down by hours.

2. Find a large piece of material with a low thread count. A natural fiber pillowcase, sheet, or cheesecloth all work, but they must not be tightly woven, or the oils in the acorns will clog them up and the water won't be able to drain through.

3. Prepare a large amount of water (5 gallons or so). Slightly warm water (not hot) will leach faster, but cold water works just as well.

4. Find a medium-sized cedar bough for the "water break."

ACORN FLOUR: LEACHING

Place your pan (which is now riddled with holes) in a place where it can have water drain through it. A really large sink, the bathtub, outside in the grass—these all work, but make sure your pan is flat. Drape the cloth across the pan and then spread the dry acorn flour across the top of the cloth in a layer that is no more than 1-inch thick.

Lay the cedar bough somewhere on top of the evenly spread flour and using a pitcher, pour the water slowly onto the bough. This bough acts as a "water break," taking the force of the water and letting the water disperse outwards across the acorn flour, drain through the flour, and drip onto the ground. Without a "water break," the force of the water will push aside the flour. Keep pouring pitcher after pitcher of water onto the bough, until all of the water is gone. Remember that different acorn species require different amounts of leaching time. Taste the flour. Bitter? You have to leach it more. Not bitter? You're ready to use it in your recipe.

Note: Don't stir the acorn flour when leaching it. That encourages the oils to separate out and the flour loses its full nutritional content.

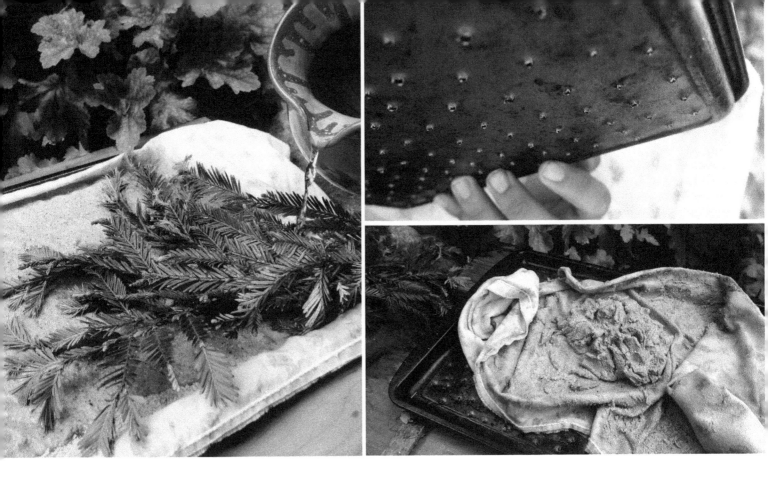

STORAGE:

DRY ACORN FLOUR CAN BE STORED IN THE FRIDGE OR THE FREEZER, BUT WET ACORN IS MORE TRICKY...

FUTURE USE: After leaching, freeze the wet flour in quart-sized bags. Flatten the flour in the bag so it freezes in a thin bar. This makes it easier to defrost. **Do not refrigerate wet acorn flour:** water and the high carbohydrate content of acorn flour makes it mold quickly.

IMMEDIATE USE: After leaching, the wet flour can be incorporated into your recipe. Remember, you are introducing an ingredient that is wet, so be mindful of your baking recipe and adjust other ingredients to arrive at the correct consistency.

Note: It is possible to use a dehydrator to get leached acorn flour dry, but it's a challenging process, and unless you're really skilled, most batches of acorn will turn out moldy.

TIPS ON BAKING/COOKING

Leaching removes most tannic acid, but not all, so it's important to remember that the tannins in acorn flour will react with various metal-based cookware and darken your dough considerably. Cook or bake with stainless steel or glass to avoid this.

Acorn Bread

by Barbara Drake

INGREDIENTS

1 C	acorn flour
	SP (pg 67)
	D+P (pg 61)
½ C	cornmeal
½ C	whole wheat flour
1 tsp	salt
1 T	baking powder
¼ C	cup honey
1	egg
1 C	milk
3 T	olive oil (or other healthy oil)

PREPARATION

Preheat the oven to 350°F. In a medium-sized plastic, wooden, stainless steel, or glass bowl (no iron!) combine the acorn flour, cornmeal, flour, salt, and baking powder. In a separate bowl, mix the honey, egg, and milk. Then add the oil. Add the wet mixture to the dry ingredients. Stir until all dry ingredients are just moistened. Pour into a greased 8 x 8 inch pan. Bake for 20-30 minutes. The bread is done when a toothpick inserted into the center comes out clean. Press down on the center of the bread and it should bounce back.

Yields one 8-inch square loaf.

Spread each square with Rose Hip Jam (page 125).

Acorn Dumplings
with Venison Stew

by Lorene Sisquoc

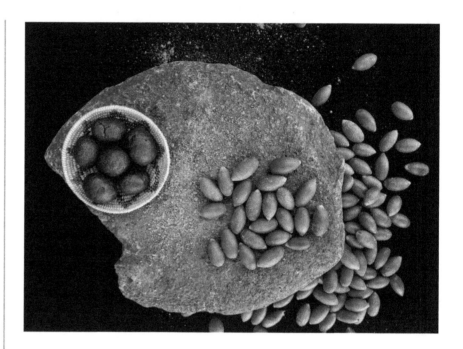

VENISON INGREDIENTS

1 lb	venison, stew-sized chunks or ground
	D+P (pg 53)
1	large onion, chopped
5	cloves garlic
👁	salt and pepper to taste

DUMPLING INGREDIENTS

½ C	acorn flour*
	SP (pg 67)
	D+P (pg 61)
½ C	whole wheat flour
1½ tsp	baking powder
2 T	milk
2 T	olive oil
1	egg, beaten

* I prefer using emory oak or gambel oak acorn flour, which comes from Arizona. It's very sweet and usually needs very little leaching—sometimes none!

PREPARATION

Sauté garlic, then add onions until they are translucent. Add ground or stew-sized chunks of venison to the garlic and onions, as well as salt and pepper to taste. Once the venison is nice and brown on all sides (with a bit of juice), add 5-6 cups of water. Let simmer on low heat for 1½ hours until the meat is tender. If using ground venison, simmer for only 20 minutes.

While the stew is simmering, make the dumplings. Mix acorn flour, whole wheat flour, and baking powder together. In a separate bowl, beat the egg and then add milk and oil to it. Combine wet and dry ingredients. Form dough into marble-sized balls (about 1½-2 inches). You may need to add 1 tablespoon of water so the dumpling balls are easier to shape. Drop the dumplings into the simmering stew. Cook for 20-30 minutes more.

GATHERING

Chia

by Deborah Small

Some writers speak of seeds not only as gifts but as jewels, an indication of how valued and revered seeds are. "Show me a seed," Henry David Thoreau wrote, "and I am prepared to expect wonders." Thoreau's persistent faith in seeds is echoed by ethnobotanist Gary Paul Nabhan, who writes that a few handfuls of native seeds contain more information than the Library of Congress. And the ethnobotanist Kathleen Harrison hears the "voices of the ancestors speaking in each of those seeds."

Here's the story of gathering California native chia seeds, *Salvia columbariae:*

"Hey, I've almost got a full ounce!" Abe announces proudly holding up his ziplock for all of us to see.

"Go for the kilo, Abe," Diania tells him.

"Yes, Abe," Marian chimes in. "Go for the kilo."

Abe Sanchez, contrary to what you might think, is not harvesting forbidden, illicit, or illegal drugs in the backcountry of North County San Diego. Along with Diania Caudell *(Luiseño),* Marian Walkingstick

(Acjachemen), Irwin Morales *(Luiseño),* and Maureen Castillo *(Cupeño),* Abe is harvesting the tiny black seeds of the chia plant. I'm along to document the harvest.

This group of basket weavers are passionately interested in anything related to traditional native basketry. In southern California, chia seeds were harvested using a seedbeater woven from sumac and other local basketry plants. The seeds were collected in a burden basket or basket tray, then winnowed to remove the chaff using a woven winnowing tray.

"For sure you want to put some gloves on," Abe tells Irwin as they're getting ready to gather. "The elders probably didn't wear them, but the chia plant has little *spiñas* on it, so if you touch it, you'll get *spiñas* in your fingers. I learned that the hard way the last time I gathered chia."

"Remember, this is rattlesnake country," Abe warns us as he descends the steep hillside from the road. It's also invasive plant country. "See this mustard right here," Abe tells us. "Pull it out! It's non-native."

I remain on the road with my video camera and tripod to get a good shot of the gathering. I'm forced to shoot through the mustard. Like so many introduced and exotic species, mustard has taken over a good deal of the land here, edging out many of the native plants. This year, it looks as if the chia is holding its own. But stands of wild chia are relatively rare.

"I'm worried about snakes," Abe says, but he doesn't seem at all worried. He's too excited about the chia seeds he's beating into a bucket with his woven seed beater.

"Look at it all down here! Look at all that!" Abe's referring to the chia accumulating in the bottom of his bucket. "You know what's good about doing this. We're dropping seed on the ground that will grow next year."

Except for Abe, no one uses any traditional tools. Irwin beats the seed heads with a pink plastic flyswatter and uses a red plastic bucket to collect them. Although his flyswatter works well enough as a seed beater, it doesn't have the beauty or resonance of Abe's woven seedbeater. And the red plastic bucket from the Home Depot, although inexpensive and functional, lacks a connection to the gathering site, unlike traditional baskets woven with the local juncus, deergrass, sumac, and *yucca*.

Irwin and Abe are in charge of gathering the chia seeds using their seed beaters on the steep slope. Marian, Diania, and Maureen are clipping seed heads in more accessible locations and placing them in a bucket.

"This is all trial and error," Abe announces. "We're learning how to do this. But it's working. It's working! Look at this! I've got a good pile already!" Everyone is too busy with their own seed collecting to look over at Abe's mounting accumulation.

Abe pulls out a particularly large mustard plant. It's important when gathering to help the habitat by pulling out invasive plants like mustard by the roots.

"How's that flyswatter working?" Abe asks Irwin.

"It's working alright," Irwin tells him. This is Irwin's first time seed beating, but he seems to be a natural. As he's beating away, I focus on the sound of swatter on seed head.

"Oh wow, you guys! I've got a good little pile in here! I've got about an ounce!!" Abe once again is all exclamation marks.

"Chia seeds are a gift," Abe tells us as he whacks the tiny black gifts into his bucket.

"We're doing an ancient thing here." Yes, we are, in spite of the pink plastic flyswatter, the red plastic bucket, and the digital video camera. We're gathering the precious wild chia seeds.

Watch the "Gathering Chia" video:

www.deborahsmall.wordpress.com/fieldtrips-2/fieldtrips-gathering-chia

Special Preparation

CHIA SEEDS

Because chia seeds do not add much flavor unless toasted, they can be added to many types of foods such as muffins, breads, and baked goods without being overpowering.

Traditionally, toasted chia is ground up and added to water to make a refreshing, thirst quenching drink. Chia use is endless, and contemporary recipes include sprinkling on yogurt, smoothies, oatmeal, and even mixed with salad dressing for an extra boost of fiber and protein. "Chia pudding" is also popular, with new recipes popping up across the internet every day. This is made with either fruit juice or with almond, soy, or other plant-based milk, and sweetened with organic agave syrup or stevia to taste. It can be enhanced with chopped fruit, raisins, currants, and spices such as cinnamon and clove.

Most of the chia seeds that are available to purchase commercially are of the *Salvia hispanica* variety, which is not indigenous to southern California. Nutritionally, it's the same as the local native variety, *Salvia columbariae*, and is used in the same ways.

If you're interested in growing chia in your garden, use raw, untreated seeds and they should sprout. The baby sprouts are tasty and good for you as well.

SOAKED CHIA SEEDS

Put the seeds in a mason jar and cover the seeds with fresh, cold water and soak at least 20 minutes. The seeds will get too thick and gooey if you let them soak too long or overnight. If soaking for baking/cooking purposes use 2:1 ratio of water to seeds. If soaking to making a beverage, use 4:1 ratio of water to seeds. This will soften the seed covering, absorb water, and make them more digestible.

STORAGE
Keep refrigerated.

TOASTED CHIA SEEDS

Chia seeds do not have much flavor themselves unless they are toasted in a dry frying pan. Toasting brings out the seeds' mild nutty flavor. Use a dry (no oil), clay skillet or cast iron pan on low heat, stir until some start to pop, and then quickly remove from heat. Don't allow the seeds to burn. You can also tell the seeds are toasted because they start to give off a unique oily scent. Spread the seeds onto a baking sheet to air-cool a half hour.

STORAGE
When toasted seeds have completely cooled, store them in a mason jar or ziplock bag. Keep refrigerated.

GROUND CHIA SEEDS

An old-fashioned, traditional rock mortar and pestle work well for small batches. Or use an electric grinder (blender, coffee grinder, etc.) if you plan to use the ground seed immediately. When using an electric grinder for large batches, be mindful that the motor will warm up and generate moisture, which can lead to eventual molding of the ground seeds if you plan to store them for a while.

STORAGE
Store in a glass jar in the refrigerator.

Creamy Chia Cacao Pudding

by Heidi Lucero

INGREDIENTS

1	14 oz. can coconut cream
1	8 oz. package cream cheese
⅓ C	organic cacao powder
1 C	almond milk, sweetened or unsweetened (can use rice, soy, cashew or regular milk as well)
1 C	chia seeds
¼ C	organic raw agave syrup
¼ C	alcohol-free vanilla flavor (Trader Joe's)

PREPARATION

Combine coconut cream, cream cheese, *cacao* powder, milk, agave syrup, and vanilla in the blender until liquefied.

In a large mixing bowl, combine liquefied ingredients with the chia. Using a whisk, stir mixture continuously until the chia seeds begin to swell. This avoids clumping. Allow to sit for 1 hour at room temperature, stirring every 10-15 minutes to avoid additional clumping. Chill and garnish with berries and chia seeds to serve!

Yields 10-12 ½ cup servings.

Chia Cornbread

by Abe Sanchez

INGREDIENTS

1 C	cornmeal
1 C	all-purpose flour
1 C	chia seeds
	D+P (pg 61)
⅓ C	sugar (optional)
2 T	baking powder
½ tsp	salt
1	egg, beaten
¼ C	coconut oil (or other healthy oil)
1 C	milk
1 C	fresh or frozen corn kernals, or canned creamed corn

I remember my mother soaking chia seeds in a jar of water and drinking the contents throughout the day. I tasted it a few times but thought it was a bit flavorless and the gelatinous seeds felt strange to swallow. As an adult, I explored other ways of eating the chia seeds I had collected from the wild. Since bread can be made with poppy seeds, I decided to do the same with chia seeds and to use a cornmeal base. Putting together these two original foods from the Americas worked! When I serve chia cornbread at Native American events, everyone always loves it and asks for more.

PREPARATION

Preheat the oven to 350°F.

In a large bowl, mix together cornmeal, flour, chia seeds, sugar, baking powder, and salt. Add egg, oil, milk, and corn. Stir gently to combine. Pour into a greased 8 x 8 inch square pan. Bake for 15 to 20 minutes or until a toothpick inserted into the cornbread comes out clean.

Chia Power Bars

by Craig Torres

INGREDIENTS

- 1 lb   *chia seeds, toasted*
  - SP (pg 79)
  - D+P (pg 61)

- 24 oz   *organic agave syrup or honey (amount depends on what else you add)*

- 👁   *coconut "powder" **

ADD SOME ZEST & CHUNK:

Toss in a variety of dried fruits and berries, like currants, raspberries, blueberries, cranberries, or cherries.

Add toasted, chopped pine nuts and/or sunflowers seeds, walnuts, pecans, or almonds.

~ Remember that if you add a variety of items, it will extend the recipe and make a larger amount. ~

....

* I find coconut powder in East Indian markets, Asian markets, or health food stores. It's just coconut, finely chopped with no sugar added. I usually toast the coconut, as it brings out the flavor.

PREPARATION

First, toast the chia (careful not to burn) and whatever other zesty and chunky ingredients (coconut, pine nuts, etc.) and then set them aside. If you toast too long or the flame is too high, the chia seeds will start to pop.

Next, you'll want to reduce the moisture content of your syrup or honey, concentrating its stickiness. Do this by pouring the syrup or honey into a small pot over medium heat, bringing it to a boil. Lower the heat and continue cooking, stirring constantly, being careful not to let it burn until you achieve a moderate thickness. Test if it's done by dropping a bit of this syrup into ice water. It should form a ball that stays firm and pliable but is still sticky between your fingers. It's important not to let the syrup get too thick, or the balls/bars may become too hard when they cool and turn out like peanut brittle.

When the syrup reaches this "ball stage," turn the heat off and pour the syrup over your dry ingredients, mixing everything together. Let the mixture cool slightly before the next step, since it will be easier to handle.

BALLS OR BARS

Balls: Roll into small, 1" or ½" balls. For an extra layer of fun, roll them in finely chopped coconut or dip them in bittersweet chocolate. The mixture is easier to handle if you periodically dip your fingers in water (but not too much water or the balls won't keep their shape).

Bars: Spread the mixture evenly onto a wax-paper lined cookie sheet that has been sprayed with non-stick cooking spray. Place a second layer of greased waxed paper over the mixture and flatten it into a solid block with an even thickness. When it cools you can score using a knife.

STORAGE

Store them in the fridge if you are making your recipe a few days or a week ahead of time. Take out and bring to room temperature before eating. This recipe can be stored in the fridge or the freezer for months. I have made it on a few occasions when the mixture did not hold up well, but don't ever think of *tossing it out;* it works great as a protein boost topping for yogurt, oatmeal, or smoothies.

This energy food can be kept in a baggie or waxed paper and used for an exercise workout boost or other outdoor activities, like hiking.

Make round balls, flat bars, or fun shapes—get creative!

Find a shady spot and spend quality time with this shake.

Chia & Mesquite
Cacao Energy Shake

by Deborah Small

INGREDIENTS

2 C	cold coconut milk (or hemp, pine nut, cashew, or any milk)
3	dates (or a few drops of stevia, or 1 frozen banana)
2 T	raw, organic cacao powder
1 T	mesquite flour *SP (pg 93) / D+P (pg 61)*
1 T	chia seeds *D+P (pg 61)*
1 T	shredded coconut
½ tsp	ground cinnamon
¼ tsp	vanilla extract
⅛ tsp	chili powder
	pinch of cloves

Chia is high in omega-3 fatty acids, and *cacao* powder is a magnesium-rich antioxidant. The mesquite, with 4 grams of fiber per tablespoon, among other nutritive benefits, makes this a super-healthy shake.

Data below is from nutritional profiles for Sunfood Raw Mesquite Powder, Sunfood *Cacao* Powder, and Nutiva Chia (organic and inexpensive at Costco). Nutritional profiles of various brands will differ slightly. The milk you choose to use will add additional fiber and protein.

FIBER

mesquite	4 grams
cacao	6 grams
chia	5 grams
total	**15 grams**

PREPARATION

Throw everything in a high-speed blender with a few ice cubes and enjoy.

Inspired by Tucson herbalist John Slattery's Sonoran Cocoa.

Special Preparation

CHOLLA CACTUS BUDS

Native people living in the southwestern United States have regularly consumed the flowering buds of various species of cholla cactus for thousand of years—a smart move considering a palm-sized dollop of cooked buds contains as much calcium as a glass of milk.

Lorene Sisquoc *(Cahuilla/Apache)* teaches her Sherman Indian High School students how to wild-gather the buds. Her ancestors gathered the fruit of the buckhorn *cholla* (*Cahuilla* name: *mutal*) and ate it fresh or dried it for storage. They also gathered the buds of the jumping *cholla* (*Cahuilla* name: *chukal*), which they cooked or steamed with hot stones in a pit, according to Saubel and Bean in their *Cahuilla* ethnobotany, *Temalpakh*.

There's nothing like gathering the buds of the spine-filled *cholla* cactus to teach respect and mindfulness. And *cholla* is a survivor. It's not just drought-tolerant. It can withstand years of prolonged drought, perfect for our changing climate. If you want to plant *cholla* in your garden, think about a safe place to site it, away from people and pets. Local plant nurseries carry *cholla*.

HARVESTING

Although "all parts of all species of *cholla* are edible," according to *Southwest* cookbook author Carolyn Niethammer, the buds are gathered in the spring when the petals are "well-formed but still tightly furled." Everyone recommends finding *cholla* with the plumpest buds and fewest spines. The buds are small, but pack a nutrient punch. They're great for lactose-intolerant folks and people working to improve their bone health. When picking by hand, wear thick gloves to avoid the spines and use kitchen tongs to twist off the flower buds from the cactus arms.

CLEAN A CHOLLA BUD OF SPINES

Wearing rubber gloves, place *cholla* buds onto a ¼-inch mesh screen and brush with a dedicated whisk broom to remove spines. Examine each bud and remove excess spines with tweezers.

PRE-PREP

We like to buy ours dried (see Description & Purchasing, page 61) so we can use them any time of year. Reconstitute the dried buds by soaking them overnight.

COOKING

Put the fresh or reconstituted buds in a saucepan, cover with water, and then simmer for 20-30 minutes until soft. Make sure to add additional water as needed. It's easy to burn them, so be careful. A ½ cup of dried *cholla* buds will yield about 1½ cups of cooked buds when reconstituted/cooked.

STORAGE

Dried buds will store indefinitely in a glass jar.
Reconstituted buds will last in the fridge 3-5 days.

Cholla Bud Succotash

by Lorene Sisquoc

INGREDIENTS

- ½ C cholla buds, dried
  - SP (pg 89)
  - D+P (pg 61)
- 2 C water
- 1 C fresh, non-GMO ears of corn (or 2 cups frozen or canned corn)
- 1 C onion, chopped
- 3 garlic cloves (or to taste), chopped fine
- olive oil (or other healthy oil)
- salt and pepper

I purchase *cholla* buds from the *Tohono O'odham* people in Arizona, but you can gather them here in California as well. You get the thorns off by rubbing them in the sand, or whatever method you use to get thorns off cactus. Then dry them to be used in future recipes!

PREPARATION

Place dried *cholla* buds in a pot and add 2 cups of water. Bring to a boil and then reduce heat, cover, and simmer about one hour or until soft. Remove from stove and drain off excess water.

If using fresh corn: Remove kernels from the cob. Stir-fry onions in olive oil about 3 or 4 minutes until translucent. Add garlic and fresh corn. Cook until corn is soft.

If using frozen or canned corn: Stir-fry onions in olive oil about 3 or 4 minutes until translucent. Add garlic. Stir-fry for another 1 or 2 minutes. Add corn and cook until corn is heated.

Add pre-cooked, drained *cholla* buds and continue to stir-fry until *cholla* buds are thoroughly heated and combined with other ingredients. Season with salt and pepper to taste.

Yields 4 cups (6 to 8 servings).

Special Preparation

MESQUITE FLOUR

As a traditional food source, mesquite provides edible parts at different times of the year. In the spring, the sticky blossoms are eaten as a sweet treat. In late spring/early summer the green pods can be eaten as a vegetable, much like string beans. In late summer, the dried pods are gathered and ground into a meal or flour for use in beverages, to make flat cakes, and gruel.

Traditionally, the *Cahuilla* used a cottonwood tree stump mortar with heavy stone pestles to break up the pods and grind them into a useable form. The seeds had to be separated from the flour by winnowing, or by soaking the flour and seeds for future use. Mesquite pods were stored in large woven willow granaries that repelled pests and provided for long-term storage.

Mesquite is quite versatile and can be used as a food or as a hot or cold tea-like beverage. We often use mesquite as a condiment or spice, in sauces, gravies, salad dressings, toppings, soups, casseroles, vegetable and meat dishes, smoothies, tortillas, and crackers.

The natural sweetness of mesquite pods and flour reduces the need to add sugar and other sweeteners in recipes. Mesquite's natural sweetness comes from the *galactomannan polysaccharide* found in the plant.

MESQUITE USED FOR BAKING OR TORTILLAS

Mesquite doesn't have gluten, which gives the elasticity to dough and allows leavening. So we combine the mesquite flour with another type of flour that has gluten. Use a ratio of ⅓ cup mesquite flour to ⅔ cup of another flour.

MESQUITE USED FOR CRACKERS (OR OTHER FLAT BREADS)

You do not need to mix mesquite flour with another type of flour.

PRE-PREP

Prior to grinding the pods, remove the thorns on the one end of the pod and stems from the other end. Then snap the dried pods into pieces about 1-inch long. It takes about one hour to prepare 1 gallon of pods to get them ready to grind. It's a good activity to do while watching a movie or chatting with friends.

If your pods are not completely dry when you grind them, they can be toasted in the oven at 250 °F for about 20 minutes to thoroughly dry the pods.

GRINDING

Today, grinding dried mesquite pods is much less labor intensive and can be done using modern kitchen appliances, such as a blender with a grinder function. In a blender, grind about 1 cup of broken pod pieces at a time, and then sift out the hard seeds with a colander.

Grinding 1 pound of flour in a blender takes 30-40 minutes. 1 gallon of pods yields about 1 pound of flour. Depending on the intended use of the flour, grind it a second time for a finer textured flour.

STORAGE

It's important that there is no moisture in the pods when you store them. Otherwise, they will mildew and spoil. Store ground flour in the freezer in air-tight containers.

Mesquite Tortillas

by Abe Sanchez

INGREDIENTS

- 1½ C unbleached flour
- 1 C mesquite flour
  - SP (pg 93)
  - D+P (pg 61)
- ½ tsp salt
- 3 T healthy oil
- ½ C warm water

The flour from honey mesquite pods has a sweet molasses-like flavor. In the Sonoran village of Desemboque in Mexico, the Seri Indians prepare it as a thick soup, which is tasty and filling. Eating it in the traditional manner is a little too rich for most of my friends. But mixing mesquite with other flours is delicious, and adding it to flour tortillas is scrumptious!

PREPARATION

In a large bowl, mix together flour, mesquite, and salt. Drizzle in oil and mix with a fork. Stir in warm water and form into a big ball. Knead for 2 minutes on a floured board. Cover with a dry, clean dishcloth and let rest 20 minutes. Divide *masa* (dough) into 12 balls and roll out into a circle. Cook in a dry skillet over medium heat. When slightly browned on one side (about 1 minute), flip and cook 15 seconds more. Remove, stack, and cover with the dishcloth. Eat immediately with butter or rose hip jam, or store at room temperature for 2 to 3 days. For longer storage, refrigerate.

Recipe inspired by the San Pedro Mesquite Company in Desert Harvesters' *EAT Mesquite* cookbook.

Wrap them around your favorite savory or sweet dish.

Mesquite & Chia
Crispy Crackers

by Deborah Small

INGREDIENTS

- ½ C mesquite flour
  - SP (pg 93)
  - D+P (pg 61)
- ½ C almond flour
- 2 T chia seeds
- 1 T coconut oil
- 1 egg
- ¼ tsp salt

They're so delicious that three friends can easily demolish the entire batch within an hour. Keep plenty of mesquite in your kitchen!

This recipe is gluten free, protein rich, high in fiber, and has a low glycemic load.

PREPARATION

Mix all the ingredients to form a stiff dough. Chill 45 minutes in the fridge. Preheat oven to 375°F. Roll dough between two sheets of wax paper to a ⅛ inch thickness. Remove the top layer of wax paper. With a knife, slice the dough into squares or triangles. Place on a foil-lined cookie sheet. Bake for 6-8 minutes, watching closely because they will burn fast!

Makes 1 to 2 dozen crackers, depending on the size.

VARIATIONS

Add a pinch of herbs to taste: Cleveland sage, black sage, desert oregano, or rosemary.

If you don't have almond flour on hand, try using cashews or another type of nut ground in a coffee grinder.

Recipe inspired by Dancy Blue French's recipe in *EAT Mesquite*, by Desert Harvesters.

Mesquite & Chia
Sage Pancakes

by Abe Sanchez

DRY PANCAKE MIX INGREDIENTS

- 2 C mesquite flour
  - SP (pg 93)
  - D+P (pg 61)
- 2 C whole-wheat pastry flour
- ½ C chia seeds
- 1 T baking powder
- 1½ tsp baking soda
- ½ tsp salt

STORAGE

Make as much dry mix as you like. Store in a cool dry place for later use.

PANCAKE INGREDIENTS

- 1 egg
- 2 T healthy cooking oil
- 2 C milk (or buttermilk)
- 2 C dry pancake mix (see above)
- 👁 fresh, whole, native sage leaves (most aromatic species will work)

PREPARATION

Whisk together the egg, oil and milk. Stir in dry pancake mix. Add water as needed to achieve the desired consistency. Ladle ¼ cup onto a lightly oiled griddle on medium-low heat. Mesquite burns easily so be careful. Lower the temperature of the griddle as needed. Watch the bottom of the pancakes for the desired brown color. Before flipping, drop a whole fresh sage leaf on uncooked side, flip and cook through. Serve with organic agave nectar, prickly pear syrup, mesquite honey, or rose hip jam.

Recipe inspired by the Desert Harvesters' *Eat Mesquite* cookbook.

Top these bite-sized pancakes with a single sage leaf or fresh fruit.

Native Superfood
Cookie Trio: Mesquite, Chia, & Pine Nuts

by Leslie Mouriquand

This cookie recipe incorporates three native superfoods: mesquite, pine nuts, and chia seeds to create a snack that is protein-rich and high fiber, with a relatively low glycemic load.

INGREDIENTS

2 C	wheat baking flour (or almond, coconut, or rice flour)
1 C	mesquite flour SP (pg 93) D+P (pg 61)
¼ C	pine nuts, chopped (or other nuts)
⅛ C	chia seeds, slightly ground SP (pg 79) D+P (pg 61)
2 tsps	baking soda
¼ C	butter (or almond butter, or ½ a banana as a low fat option
½ C	shortening (or almond butter, or ½ banana as a low fat option)
⅓ C	brown sugar (or coconut palm sugar or organic agave sweetener)
2	eggs
1 T	vanilla extract (or more to taste)

PREPARATION

Set oven to 350°F. In a large bowl, mix together wheat flour, mesquite flour, pine nuts, chia seeds, and baking soda. In a separate bowl, blend butter, shortening, brown sugar, banana, eggs and vanilla. Mix dry and wet ingredients together. Flatten cookie balls with a fork to bake more thoroughly. Bake on an ungreased (or parchment lined) cookie sheet for 13 to 14 minutes.

Yields 24 to 36 cookies.

It's true... cookies can be healthy!

Special Preparation

PRICKLY PEAR CACTUS

There are over 200 species of cactus worldwide, but not all are edible. Prickly pear cactus are edible and the best tasting ones are found in the southwestern United States (although some species grow as far east as Massachusetts). Prickly pear cactus produces two parts which are used in cooking: the pads and the fruit.

Prickly pear can survive in ferocious conditions, including our wildfires. Prickly pear is also the ultimate sustainable plant. As our friend Willie Pink *(Cupeño)* says, "gather a pad from a wild prickly pear, and two will grow in its place." It's easy to propagate and most importantly, requires no supplemental irrigation, ideal for our water-challenged region. To propagate, cut a pad and wait two weeks for the wound where you severed it from the mother plant to heal over. Then plant the pad you cut, and in two years you'll have fresh young pads and tunas to eat. You can plant it among *agave, yucca,* or *dudleya.* The bees, birds, and other wildlife will thank you.

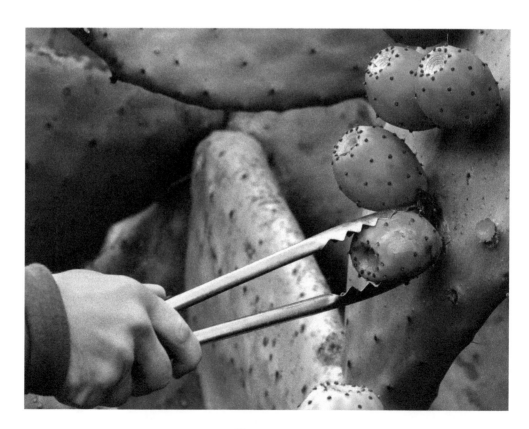

CACTUS PADS (NOPALES)

Used in Native cuisine for hundreds of generations, *nopales* are prepared in various ways: cut up and boiled, grilled, roasted, baked, or stir fried. They can be added to scrambled eggs, chili-based dishes, salads (cooked), or pureed to make drinks. *Nopales* are available year-round but are best from early spring through late fall. When buying *nopales,* choose small (6-8 inch), firm, pale green pads with no wrinkling. Be sure to choose pads that are not limp or dry. When picking by hand avoid the sharp glochids (hairs) by wearing thick gloves.

CLEANING A CACTUS PAD OF ITS GLOCHIDS

While wearing gloves, place a fresh cactus pad on a cutting board. Holding the pad from its trunk base, use a knife to shave the glochids away with upward strokes until they are all removed. Turn the pad over and repeat the same process on the other side. Trim off the outer edges of the pad with the knife to remove all remaining glochids.

STORAGE

Use immediately in your recipe, or place in a paper bag and store in the refrigerator until needed. You can also freeze or can the pads.

COOKED CACTUS PADS

Nopales exudes a gelatinous liquid when it's cooked. We like to preserve as much of this nutritious gel as possible. Grill them or sear them in a skillet with a bit of oil for just a few minutes on each side to preserve the liquid. Or boil them for a short time in water, (the water you've boiled them in can be used in the recipe as well).

BLENDED CACTUS PADS

Combine a small amount of water and cactus pads (cut to fit in blender), and blend until liquefied.

STORAGE

Use immediately in your recipe, or freeze in storage containers.

Special Preparation

PRICKLY PEAR CACTUS

CACTUS FRUIT (TUNAS)

Fresh *tunas* are available in the fall and can be dark red, magenta, or apricot colored. When harvesting, wear thick gloves to avoid the sharp glochids (hairs) and use kitchen tongs to twist off the fruits from the pads.

CLEAN A CACTUS FRUIT OF ITS GLOCHIDS

Wear rubber gloves and clean the fruits under running water using a kitchen scrubber to rub off all spines and glochids. If a fruit has longer spines, hold it with tongs and burn off the glochids and spines on a gas stove. Many people do this outside using a camping stove.

STORAGE

Use immediately in your recipe, or place in a paper bag and store in the refrigerator until needed.

JUICED CACTUS FRUIT

After removing the spines and glochids, cut off the top and bottom of each fruit. Put the whole fruits through your juicer until you produce enough juice for your recipe. Toss, or better yet, compost the pulp and seeds.

STORAGE

Use immediately in your recipe, or freeze juice in ziplock bags or ice cube trays.

Nopales Salad

by Abe Sanchez

INGREDIENTS

- 4 handfuls Roma tomatoes, chopped
- 1 fresh jalapeño or serrano chile, chopped
- 1 C onions, chopped
- 1 can pinto beans, drained (or fresh cooked beans)
- 1-2 cactus pads, cooked and chopped
  - SP (pg 103)
  - D+P (pg 61)
- 1 bunch cilantro, chopped
- 👁 lemon juice (about 1 lemon)
- 👁 salt to taste
- 1 C queso fresco (or feta cheese) (optional)

PREPARATION

In a large bowl, combine the tomatoes, *chile,* onions, drained beans, *cilantro,* lemon juice, and cactus (which has been cooked, chopped and cooled). Salt to taste. Toss to mix evenly. Serve salad topped with crumbled *queso fresco.*

Nopales Stir Fry

by Lorene Sisquoc

INGREDIENTS

¾ C   onions, chopped

3   garlic cloves (or to taste), chopped fine

1 C   non-GMO corn (fresh, frozen or canned)

2   cactus pads, cooked and chopped (makes about 1 cup)

     SP (pg 103)
     D+P (pg 61)

¼ C   olive oil (or other healthy oil)

     salt and pepper

PREPARATION

If using fresh corn: Remove kernels from the cob. Stir-fry onions in olive oil about 3 or 4 minutes until translucent. Add garlic and fresh corn. Cook until corn is soft. Add the cooked and chopped *nopales* (cactus pads). Stir until mixture is heated.

If using frozen or canned corn: Stir-fry onions in olive oil about 3 or 4 minutes until translucent. Add garlic. Stir-fry for another 1 or 2 minutes. Then add the corn and cooked nopales. Stir until mixture is heated.

Season with salt and pepper to taste.

Nopales Tortillas

by Abe Sanchez

INGREDIENTS

1	large cactus pad, blended until liquified (makes about ½ - ¾ cup)
	SP (pg 103)
	D+P (pg 61)
½ C	water
1 C	masa harina
	D+P (pg 61)
½ tsp	salt

DOUBLING THE RECIPE

Make sure the cactus is liquified well enough so that the ball of dough doesn't develop cracks when kneading. Don't press the tortillas too thin or they will be difficult to get off the plastic.

When I was in Oaxaca, I noticed vendors were selling many different kinds of corn tortillas. I assumed that these creative ways of preparing the tortillas must enhance the locals' daily corn tortilla consumption. Fresh corn tortillas are delicious, but when you eat them every day, a little change in flavor can make meals a lot more exciting. By mixing *masa harina* with greens—*nopal,* lamb's quarters, nettle, or amaranth, you have two vegetables in one, with enhanced flavor.

PREPARATION

Blend the *nopal* (cactus pad) in water until liquified. In a separate bowl, mix the *masa harina* and salt. Slowly stir in the blended *nopal* mixture until the dough can be shaped into a ball. Knead a few times by hand. Form into a large ball and let sit for 30 minutes. Divide dough into 8 sections and roll into small balls.

Cut a gallon-sized freezer bag around the edges to create 2 sheets of plastic. Place a ball of dough between the plastic sheets and use a rolling pin to flatten it into a circle. Being careful of the edges of the tortilla dough, slowly peel back the plastic sheets.

Drop into a hot non-stick or cast iron pan. Let cook for one minute. Flip and cook one minute more. The tortilla should start to puff up while cooking. If it doesn't, flip it one more time.

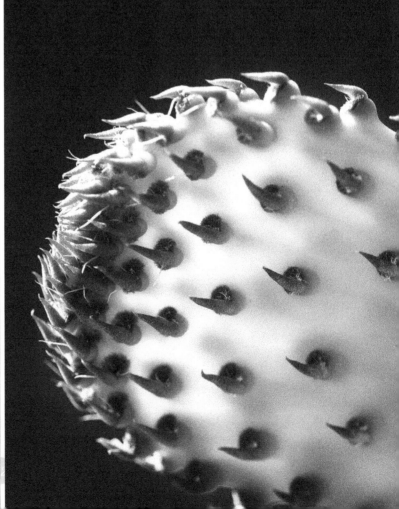

Special Preparation

TEPARY BEANS

"By using our ancestral foods, we can reverse Type-2 diabetes, we can reverse heart disease; we can get people off their meds," Kiowa chef and writer Lois Ellen Frank tells us, "but we have to know how to prepare the foods. That's how important traditional ecological knowledge is. Scientists are predicting that this is the bean that can save the world."

One of the most drought resistant crops anywhere, the tepary bean has been grown in the Sonoran desert area with minimal irrigation for thousands of years. In their *Cahuilla* ethnobotany book, **Temalpakh**, Saubel and Bean write that tepary beans *(tevinymalem)* were cultivated by the *Cahuilla* as part of their trade economy.

According to anthropologist Mike Wilken, elders of the *Paipai* tribe in Santa Catarina, Baja California, Mexico remember eating tepary beans as children, which they called *yurimun*. Their late elder Benito Peralta recalled that they ate white, brown, and other colors of cultivated tepary beans.

According to Native Seeds/SEARCH, the normally high yield tepary beans have been documented to produce lower yields with increasing amounts of water. Their roots grow twice as deep as other beans and their growing time is shorter.

The high-protein, high-fiber tepary beans are high in soluble fiber, low on the glycemic index (29) and can help regulate blood sugar levels. The nutrient dense beans are also higher in calcium than other beans. In our unpredictable, climate changing world, we all should pay attention to this little bean whose adaptations have enabled it to survive fierce heat and drought conditions.

You can try growing your own tepary beans. In their effort to support seed sovereignty and food security, Native Seeds/SEARCH sells 33 varieties of wild and cultivated tepary bean seeds.

PRE-PREP

Tepary beans are smaller but denser than other beans, and they take longer to cook. First, sort through the beans to remove any pebbles or dirt. Wash thoroughly and then presoak overnight before cooking (both white and brown tepary beans have the same soak time). We bring a big pot of water to a boil, turn off, and add the beans. This is really important if you've stored your beans for a long time.

COOKING

In the morning, drain them, then place them back in the pot. Cover them with fresh water, bring to a boil, and simmer for 2-4 hours until tender. The beans can also can be cooked in a crockpot. Wait until they are fully cooked before adding salt and seasonings. The beans double in size after cooking.

STORAGE

They store well in the fridge for 5 days, or can be frozen for longer storage.

Nopales
Tepary Bean Salad

by Lorene Sisquoc

INGREDIENTS

- 1 C — dry, white tepary beans
  - SP (pg 113)
  - D+P (pg 63)
- 1 C — dry, brown tepary beans
  - SP (pg 113)
  - D+P (pg 63)
- 4 — fresh, non-GMO ears of corn (or 2 cups frozen or canned corn)
- 1 C — onion, chopped
- 3 — garlic cloves (or to taste), chopped fine
- 2 T — olive oil (or other healthy oil)
- 1 C — cactus pads, cooked and chopped
  - SP (pg 103)
  - D+P (pg 61)
- ½ C — black beans, cooked or canned
- ½ C — kidney beans, cooked or canned
- ½ C — pinto beans, cooked or canned
- salt and pepper

I was introduced to tepary beans years ago through the *Tohono O'odham* people. I love them. I even grew some in my front yard. I like to use them because they're one of the traditional beans that is still used today. They're a great diabetic food, high in fiber. I remember that [the late *Cahuilla* elder] Alvino Siva said we had our own traditional beans here in California. He even described what they looked like. He said you could sometimes find them in the wild, but I've never seen them.

PREPARATION

If using fresh corn: Remove kernels from the cob. Stir-fry onions in olive oil about 3 or 4 minutes until translucent. Add garlic and fresh corn. Cook until corn is soft. Chop cooked nopales into small cubes and add. Stir until mixture is heated.

If using frozen or canned corn: Stir-fry onions in olive oil about 3 or 4 minutes until translucent. Add garlic. Stir-fry for another 1 or 2 minutes. Then add the corn and cooked nopales. Stir until mixture is heated.

In a large bowl, mix the tepary, black, kidney, and pinto beans. Then add the stir-fried onions, garlic, corn, and cooked and chopped *nopales* (cactus pads). Toss together using the olive oil they cooked in as the dressing. Season with salt and pepper to taste.

Yields 6 cups (6 to 12 servings).

High in fiber and protein, this bean regulates blood sugar.

Tepary Tart & Fruit Compote

by Craig Torres

INGREDIENTS

- ½ C mesquite flour
- 1 C coconut flour
- 1 C almond flour
- ½ C coconut oil (solid)

THE CRUST

PREPARATION

Mix flours together evenly. Add coconut oil and blend with your hands until it resembles graham cracker crust. Pour into a 9" springform pan and use the bottom of a rounded measuring cup to compact the crust into an even thickness. Set aside and refrigerate.

THE FRUIT COMPOTE

PREPARATION

Combine your choice of berries—blueberries, elderberries, blackberries, raspberries, mixed berries, or cactus puree—in a saucepan with 1/2 cup of water or juice. Cook for 5-10 minutes, until bubbling. Lightly smash the berries with the back of a spoon or fork. Remove saucepan from heat.

Mix arrowroot powder in a small bowl with 3 T of water. Add slowly to berry mixture and stir constantly until dissolved. As mixture cools, add 1 T of honey or sweetener of your choice. Set aside.

INGREDIENTS

- 2-3 C *fresh or frozen berries, or prickly pear cactus puree*
- 2 T *arrowroot powder (or substitute cornstarch)*
- ½ C *water*
- 1 T *honey or sweetener of your choice*

INGREDIENTS

2 C	dry, white or brown tepary beans
	SP (pg 113)
	D+P (pg 65)
2 C	tepary bean water (saved from cooking the beans)
3 T	lemon juice
3 T	alcohol-free vanilla flavor (Trader Joe's)
3 T	gelatin, sustainably sourced
¾ C	water
¾ C	organic coconut palm sugar (Trader Joe's)

Note: If you don't have a springform pan, you can use a baking pan.

THE FILLING

PRE-PREP

Tepary beans take the longest of all beans to cook. To shorten cooking time, bring your beans to a boil in about 5 quarts of water, then turn off and let sit in its water for two hours. Light up the fire again and cook for 2-4 hours, or until done. Pinch a bean... if it's soft, you're done! DO NOT ADD SALT, because it will actually extend your cooking time. 2 cups of dried beans will yield approximately 6 cups of cooked beans.

PREPARATION

After straining your pre-cooked tepary beans (save the water!), puree them in a blender with enough bean water (about 2 cups) to create a smooth, creamy consistency. Add the lemon juice, vanilla, and coconut palm sugar (adjust to your taste), and blend for another minute or two.

Combine 3 T gelatin in ¾ cup of cold water in a saucepan. Let stand for 5-10 minutes. Then heat gently and stir until dissolved. Add to blended bean mixture and blend for another minute or so.

Pour into the springform pan on top of the crust. Chill in the refrigerator for 2-8 hours, until it sets into a flan-like consistency. Take out of fridge, remove from springform pan, pour and spread berry compote over the tart, and serve to the vultures (er, friends).

Prickly Pear Cactus
Frozen Treats

by Leslie Mouriquand

INGREDIENTS

- 3 C   cactus fruit juice
  - SP (pg 105)
  - D+P (pg 6)
- 1 C   diet Sprite, 7 Up, or Fresca
- 2 T   chia seeds, soaked
  - SP (pg 79)
  - D+P (pg 6)

VARIATIONS

WILD GRAPE FROZEN TREATS

- 3 C   wild grape juice
- 1 C   diet Sprite, 7 Up, or Fresca
- 2 T   chia seeds, soaked
- ½ C   fresh blueberries (optional)

CHIA LEMONADE FROZEN TREATS

- 3 C   fresh squeezed lemon juice
- 1 C   diet Sprite, 7 Up, or Fresca
- 2 T   chia seeds, soaked
-    small chunks of fresh banana (optional)

PREPARATION

Combine all ingredients together and pour into popsicle molds or plastic disposable cups. You can eliminate the soda pop and add another cup of juice for an all-natural treat. Insert popsicle sticks through the bottom center of a cupcake liner for each one and turn the liner upside down over the cup like a lid so that the stick stays upright and centered while freezing. Freeze for at least 10 hours. Remove the cupcake liner and enjoy.

The ratio used above is 3:1, which yields 4 cups. You can adjust quantities for the size and number of popsicle molds you are using. Use 4-ounce popsicle molds for 8 servings or 2-ounce plastic disposable cup molds for 16 servings. I often double and triple the amounts of juice, soda, and chia seeds for large groups.

Prickly Pear
Marinated Quail with Mesquite Rub

by Abe Sanchez

This recipe became one of our favorite dishes prepared by students each summer at UCLA's Pimu (Catalina Island) Archaeology Field School. The students cook native foods in an outdoor kitchen overlooking the ocean. To avoid splattering and to seal in the flavor of the prickly pear marinade, we place the butterflied quail in the sun to dry them out before frying. We squeeze lemon juice on the quail to deter yellow jackets from landing on the birds.

INGREDIENTS

- 6 — dressed quail, whole or butterflied (or chicken or any firm fish)
- 2 — lemon or limes
- salt and pepper
- 2 C — cactus fruit juice
  - SP (pg 105)
  - D+P (pg 61)
- 1 C — mesquite flour
  - SP (pg 93)
  - D+P (pg 61)
- ¼ C — olive oil
- coconut oil (or other healthy cooking oil for frying)

PREPARATION

Squeeze lemon juice over birds. Salt and pepper them. Marinate in the prickly pear juice for 2 hours or overnight in the refrigerator. The dark red or orange-colored juice will stain the raw meat a beautiful color, but the cooked meat will not retain the color. Remove birds from marinade and pat dry. Brush with olive oil. Rub with mesquite flour. Bake or fry.

TO BAKE

Preheat oven to 350°F. Lay birds on a baking sheet and bake for 45 minutes or until brown and crispy. Butterflied birds will cook faster than whole birds.

TO FRY

Place in hot cooking oil, and turn over until golden brown.

Pair with Nopales Tepary Bean Salad (page 115).

Rose Hip Jam

by Barbara Drake & Deborah Small

INGREDIENTS

¼ C rose hip powder
   D+P (pg 63)

¾ C cactus fruit juice (or use unfiltered organic apple juice)
   SP (pg 105)
   D+P (pg 61)

1 T chia seeds
   SP (pg 79)
   D+P (pg 61)

VARIATION

Add lemon, lime, or orange peel powder to enhance the vitamin C content.

Barbara: The rose hip itself is the medicine. It's a really good tea to drink when you start to catch a cold because it's full of vitamin C. The tea is also great for soothing sunburned skin or for minor kitchen burns.

PREPARATION

In a bowl containing the rose hip powder, slowly add prickly pear juice until the mixture is thick and creamy. Add chia seeds and additional prickly pear juice. Stir and let stand for 15 minutes while the chia thickens the mixture. Check the consistency to see if more juice is needed. The pectin in the juice helps bind the jam. Transfer to two or three 4-ounce jars. You can refrigerate for about a week, or freeze for longer storage.

Makes about one cup of jam.

This recipe was inspired by Valerie Segrest at the 2013 TOCA Native Foods Symposium in Tucson, Arizona.

Load up a Mesquite Pancake (page 99) with this tart treat.

Special Preparation

STINGING NETTLE

FRESH STINGING NETTLES

Wild nettles can be found in your yard or vacant lots. Edible nettles are ones which have not been sprayed with pesticides or herbicides and are also not growing close to the highway (exposed to toxins). When harvesting nettles for food or tea, get the leaves before the plant flowers. New at this? **Wear gloves to gather.** Cut the tender tops (usually 4-6 leaves or 2-3 leaf sets). It's best to wear a long-sleeved shirt and long pants when gathering.

STORAGE

Freshly harvested nettles will store in a bag in the vegetable drawer of the refrigerator, unwashed, for at least 2-3 days.

COOKING

Blanching with boiling water takes out their sting and then the nettles can be added to your recipes. Boiling them in soups will also eliminate the sting. Blanched nettle leaves can be sautéed, steamed, or juiced.

DRY STINGING NETTLES

Dried leaves can be used as tea or ground up in a spice grinder and added as a supplement boost to smoothies for a healthy dose of vitamins, minerals and chlorophyll. Stinging nettles are utilized medicinally in the form of hot and cold infused tea and tinctures. When used on a regular basis, check with your doctor about possible interactions with prescription medication. Some online herbal stores offer dried nettle leaves for sale (see **Description & Purchasing, page 63**).

STORAGE

Keep in a glass jar out of the sun.

USE

Depending on the recipe, use the leaves whole or grind them with a mortar and pestle.

Stinging Nettle
& Sunflower Seed Soup

by Barbara Drake

INGREDIENTS

- 3 garlic cloves, minced
- 1 medium onion, diced
- 3 T olive oil
- 2 qt low sodium chicken broth, vegetable broth, or water
- 2 C raw, unsalted, shelled sunflower seeds
- 1 herb bundle tied with string (3-inch sprig each of thyme, sage, and rosemary and 1 bay leaf, fresh or dried)
- 2 handfuls fresh stinging nettles (or 1 cup dried)

SP (pg 127)
D+P (pg 63)

PREPARATION

Always use gloves or tongs when handling fresh nettles. In a soup pot, sauté garlic and onion in olive oil. Add broth or water, sunflower seeds, and the herb bundle. Simmer about 20 minutes until sunflower seeds soften. Add nettles and simmer for an additional 10-15 minutes. Remove herb bundle and serve.

Serves 6-8.

Nutritional information: nettle is rich in iron, magnesium, calcium, boron, and carotenoids

Stinging Nettle
Tea Medley

by Barbara Drake

INGREDIENTS

| 1 part dried stinging nettle leaves
  SP (pg 127)
  D+P (pg 63)

| 1 part dried rose hip powder
  D+P (pg 63)

| 1 part dried mint leaves

 honey (optional)

PREPARATION

Combine all dry ingredients in a mason jar. Measure 1 teaspoon of the dried tea mixture for each 8-ounce cup. Pour boiling water over the dried tea and let steep 2-3 minutes (or longer for a stronger tea). Strain out leaves before serving. Sweeten with honey if desired.

The Chia Café Collective makes this medley for our workshops, classes, and demonstrations.

Enjoy as a hot or cold tea with Mesquite & Chia Crackers (page 97).

YUCCA MOTH

Dancer

by Barbara Drake: Mother, Grandmother, Great-Grandmother

This is the story of a little white moth who is a mother just like many of us. Her life's purpose is to find a safe home among the *yucca* flowers (with their special food to feed her children).

She begins her dance as she visits the creamy white flowers of the *yucca* stalk. She lays her eggs inside a flower. She dusts the eggs with a cloud of pollen.

The tiny larvae will hatch and feed on the delicious flower seeds. Any pollen remaining will fertilize seeds and develop into a sweet, green *yucca* fruit.

The *yucca* moth dancer and the majestic *yucca* plant have a symbiotic relationship, which simply means they need each other to live.

The dance will begin again next year with *yucca* moth's children returning to the *yucca*, a dear friend.

Special Preparation

YUCCA

From a Native perspective, part of the traditional knowledge about the yucca is knowing what, where, and when to gather, and how to prepare the different parts of the plant not only for immediate consumption but for later use as well.

The *yucca* plant and its various parts are a staple for many southern California Native people. Everyone thinks in terms of blossoms, but the young stalk is also edible. If you miss the opportunity to harvest the stalk, you can go back and collect the blossoms. If you miss the blossom season, you can go back and get the fruit pods. If you miss the young, green fruit pods, you can return several weeks later to gather the dried pods, break them open, collect the seeds, and grind them down into a flour.

In many states, *yucca* is a protected plant and can only be harvested with a permit, which is why it is not available commercially. We gather blossoms on our friends' lands for workshops and gatherings. Fresh blossoms are edible in very small quantities and can be sprinkled on salads, but be careful not to eat too many, as they may irritate your throat.

PRE-PREP

If the fresh blossoms have grown large and full, twist out the center of the blossom before using, as those become increasingly bitter with age. If gathering blossoms when they first start to open, there is no need to remove the center portion.

BLANCHING

Fresh *yucca* blossoms are boiled in 2 or 3 changes of water (15-20 minutes each time) to remove a bitter yellow liquid. When the water is a pale yellow, drain blossoms and hand-squeeze to eliminate any excess moisture. The blossoms are now ready to be used in recipes. Fresh blossoms should be used within a few days.

If you air-dry or dehydrate the blossoms for later use, you still need to blanch them to remove the bitterness before reconstituting them for use in a recipe.

STORAGE

Whole blossoms can be blanched and then frozen, or frozen and then blanched. Dry or dehydrated blossoms can be stored in an air-tight container.

Marinated Yucca Blossoms

by Barbara Drake

INGREDIENTS

- 2 C yucca petals, cooked
  - SP (pg 135)
  - D+P (pg 63)
- 1 C artichoke hearts, chopped and marinated
- 1 three-inch sprig each of fresh thyme, sage, and rosemary
- 2 garlic cloves, mashed

MARINADE

- ½ C olive oil
- ¼ C rice vinegar
- 👁 zest and juice of 1 lemon
- 👁 organic, raw agave syrup (or sweetener of your choice, to taste)

PREPARATION

Place the *yucca* petals, artichoke hearts, herb sprigs, and garlic in a quart canning jar. Cover with marinade. Make sure the tops of the *yucca* petals are covered with extra olive oil. Cover and put in the refrigerator for several days until all flavors have blended.

Serve on a water cracker or toasted pita slice.

Yucca Petal Hash

by Barbara Drake

INGREDIENTS

1	onion, sliced
3	garlic cloves, minced
3 T	olive oil
½	green or red bell pepper, sliced
1 T	olive oil
2 C	yucca petals, cooked

SP (pg 135)

D+P (pg 63)

1 tsp	sugar (or sweetener of your choice)
1	tomato, chopped
½ C	cholla buds, boiled (optional)

SP (pg 89)

D+P (pg 61)

 salt and pepper to taste

PREPARATION

Sauté the onion, garlic, and green or red pepper in oil for 3 minutes. Add the *yucca* petals. Continue to sauté 10-15 more minutes until the *yucca* petals are tender but still hold a light crispness. Add sugar and stir in the chopped tomato and *cholla* buds. Heat thoroughly. Add salt and pepper.

Serve on tortillas or as a side dish.

Fill a Mesquite Tortilla (page 95) with this flavorful dish.

I Fought

TYPE-2 DIABETES AND WON WITH NATIVE FOODS

by Leslie Mouriquand

For many years I worked in an office—always sitting. With a heavy work load, I usually worked through lunch or went out to get a quick drive-through lunch, eating while I drove. Some call this multi-tasking, but I call it insanity...now. If there was a competition for driving down the freeway while eating a Del Taco Classic Chicken burrito with one hand, I would have won the prize!

After years of this, it caught up with me. I went to see my doctor for my annual blood work and was told that I had Type-2 diabetes. That scared the pants off of me! I also felt ashamed; ashamed because in my spare time I had been participating in Native food classes, workshops, and events, and I had been touting the benefits of traditional foods to all who would listen. And here I was not practicing what I was preaching. Intellectually, I knew better, but had habituated myself to the "one-handed burrito diet."

For a year I followed my doctor's advice and took my prescription pills. I hated having to check my blood sugar levels all the time and found that I struggled with regulating my blood sugar levels because of the medicine. Knowing that diabetes had reached epidemic levels in the United States and knowing what the health concerns were with this disease, it dawned on me what I needed to do: totally change what and how I ate. I needed to get off my duffer and MOVE! I needed to actually eat the traditional foods that I had been researching and talking about to others, not just for cultural preservation sake, but to save my life!

My journey began with cutting cactus pads from the wild or buying them in jars at the grocery store, gathering mesquite pods and grinding them, ordering chia seed, and starting a garden. I took classes for two years, earning a certificate as a Master Desert Gardener. I began experimenting in the kitchen and conjured up some basic recipes. Dedicating myself to a strict, traditional plant-food diet was not realistic for me since I still worked and interacted with the rest of the world. I had to figure out how to incorporate these foods into my modern diet. It had to work with my lifestyle as much as possible. I had been focusing on learning about traditional uses of honey mesquite so I came up with a mesquite cookie recipe and played with substituting sugar and butter with other, healthier ingredients. I had fallen in love with honey mesquite and went crazy playing with it in the kitchen, using my friends as my taste-testers.

Cactus was next. I scraped, burned, and used pliers to remove the glochids (thorns) on the cactus pads. Many evenings were spent tweezing the darn things out of my fingers! Jarred cactus is loaded with salt as a preservative, so I limited the amount of jarred cactus strips that I ate. I removed as much of the salt as possible by dumping them into a colander and rinsing with cold water. They are great in scrambled eggs and in salads as well as stir-fry dishes. But the fresh cactus strips are so much better for you. A good friend of mine experimented with making dried cactus strips as a snack food. I think he is on to something; they were delicious!

Knowing that part of my fight against diabetes was going to involve getting a lot more exercise, I started walking every chance I got. My sister gave me a recumbent stationary bike and I pedalled away on it every evening, gradually increasing the number of miles each week.

Realizing that I couldn't gather all the native foods that I needed to be eating from the wild, I started a garden out back. Gardening gave me another form of exercise, and a peaceful place to unwind and relax from the craziness of life. My garden, aptly named, the **'Garden of Weedin,'** is unconventional. I planted seeds, plants, and trees in and around the native plants that were already there, hoping that a symbiotic plant magic would happen. Lots of trial and failure at first, but now, I have more success in growing good food: I gather seeds from the wild and plant them, I order seeds from websites that specialize in native plants and foods, I save seeds from the plants that I grow and replant them the next year, and I give some away to others. I also keep learning about traditional food plants.

Common garden veggies and herbs are also a huge part of my garden, in addition to some exotic ones from other regions (*Hopi* red amaranth, *Hopi* blue corn, tepary beans, panic grass, chia, epazote, giant sunflower, etc.) typically grown by tribal people who live in higher altitudes. I can make one heck of a United Nations salad or stir-fry with plant representatives from all over the Southwest!

Now I am part of gatherings and events focused on traditional foods, but instead of donning my anthropologist hat, I am a committed native foodie! Passionately immersed in native foods, I am happy to report that within a year of my diabetes diagnosis and changing my diet, I was off of diabetes medication and my doctor reclassified me as non-diabetic. I had succeeded in reversing my diabetes and I was his first patient to do this. If I fall off the native foodie wagon and go back to my old unhealthy ways it could return, so this is a lifetime commitment for me.

EATING A DIET FULL OF NATIVE FOODS HAS MADE ME PASSIONATE ABOUT CONCOCTING NEW RECIPES TO SHARE WITH MY FRIENDS.

Try

1 tsp of mesquite flour in your morning oatmeal (¼ cup uncooked oatmeal made with water or chia gel) with 1 tsp of bruised chia seed, sprinkled over the top and crowned with a few banana slices or 1 tsp of almond butter. Yum! Mesquite flour and chia seed stabilize blood sugar and keep you going all morning. For a higher protein breakfast, try making an omelet with 1 tsp of chia seed, ½ cup cactus strips, ¼ cup sliced sweet peppers and mushrooms, 2 tbsp of salsa, and a little bit of a white cheese.

Now, my freezer is full of traditional foodstuffs, like mesquite flour, pinyon pine nuts, cactus puree, cactus fruit juice, *yucca* blossoms, and purslane, not just for myself, but also for all of the Native food tasting events that have become so popular among both Native and non-Native peoples.

There is so much room for culinary creativity with Native foods; they can be frozen, pickled, and dried just like more 'mainstream' foods. Everyone can benefit from incorporating traditional native foods into their diets whether they have diabetes or not.

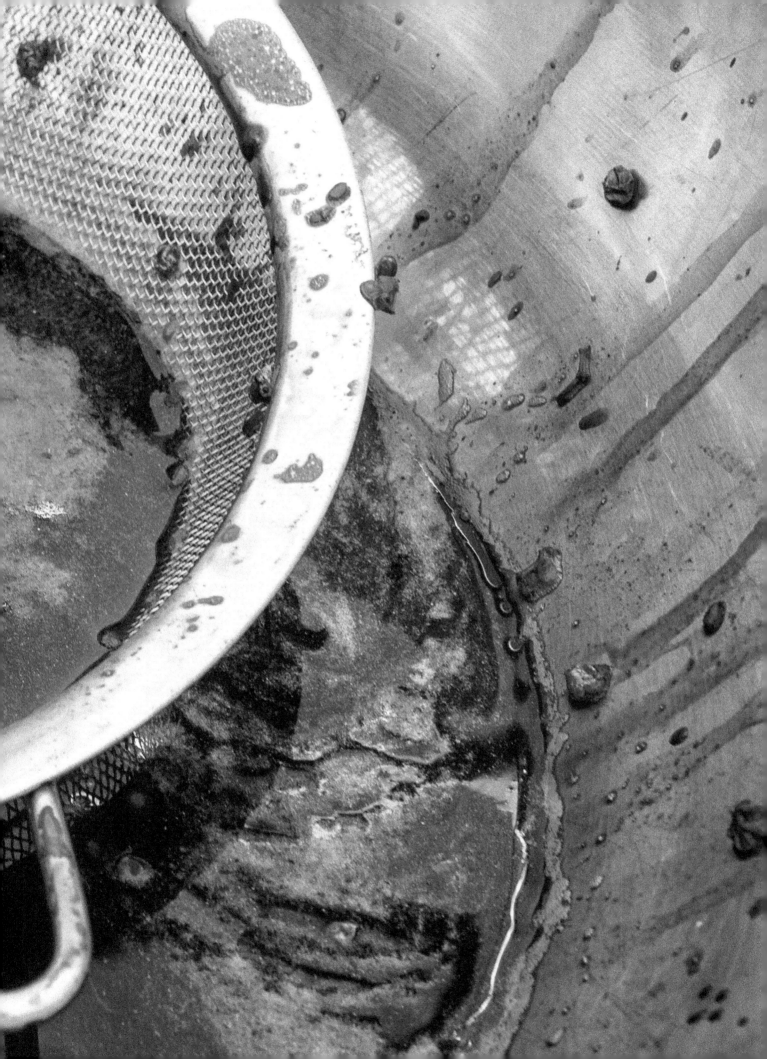

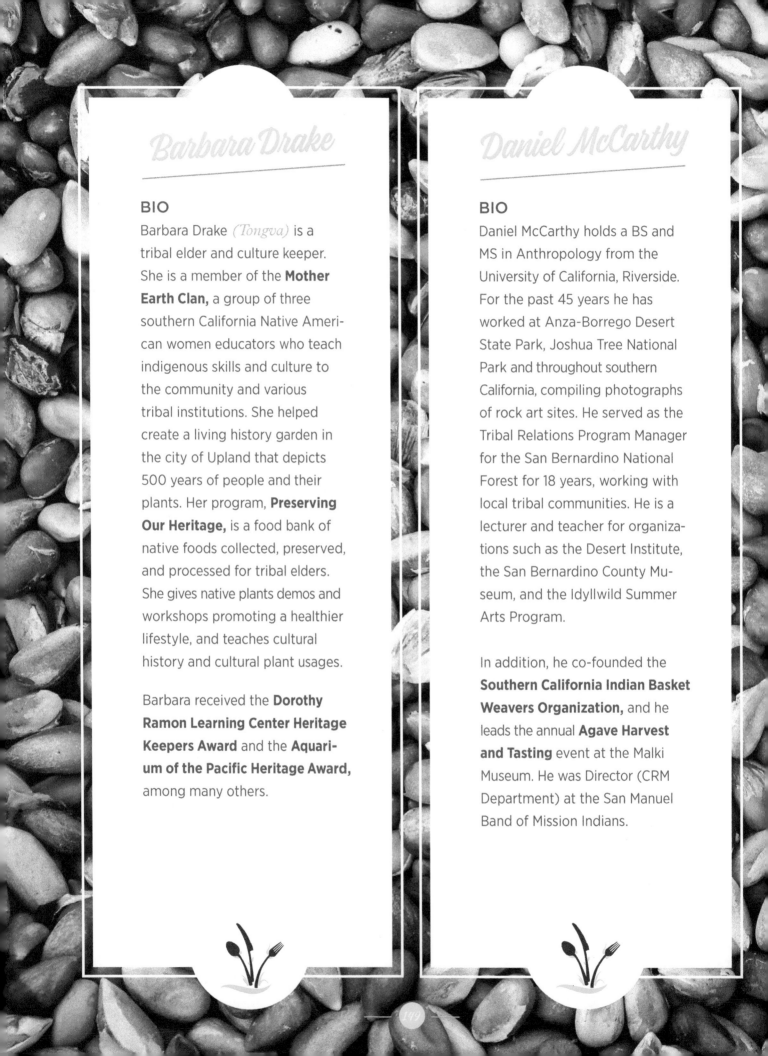

Barbara Drake

BIO

Barbara Drake *(Tongva)* is a tribal elder and culture keeper. She is a member of the **Mother Earth Clan,** a group of three southern California Native American women educators who teach indigenous skills and culture to the community and various tribal institutions. She helped create a living history garden in the city of Upland that depicts 500 years of people and their plants. Her program, **Preserving Our Heritage,** is a food bank of native foods collected, preserved, and processed for tribal elders. She gives native plants demos and workshops promoting a healthier lifestyle, and teaches cultural history and cultural plant usages.

Barbara received the **Dorothy Ramon Learning Center Heritage Keepers Award** and the **Aquarium of the Pacific Heritage Award,** among many others.

Daniel McCarthy

BIO

Daniel McCarthy holds a BS and MS in Anthropology from the University of California, Riverside. For the past 45 years he has worked at Anza-Borrego Desert State Park, Joshua Tree National Park and throughout southern California, compiling photographs of rock art sites. He served as the Tribal Relations Program Manager for the San Bernardino National Forest for 18 years, working with local tribal communities. He is a lecturer and teacher for organizations such as the Desert Institute, the San Bernardino County Museum, and the Idyllwild Summer Arts Program.

In addition, he co-founded the **Southern California Indian Basket Weavers Organization,** and he leads the annual **Agave Harvest and Tasting** event at the Malki Museum. He was Director (CRM Department) at the San Manuel Band of Mission Indians.

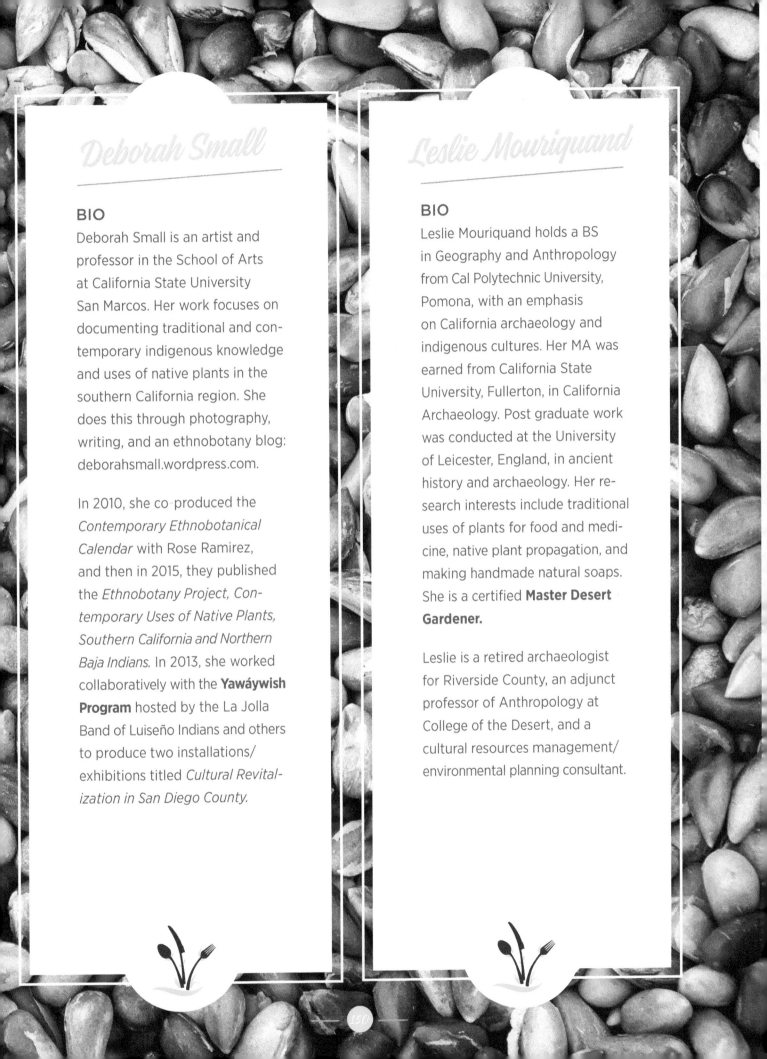

Deborah Small

BIO

Deborah Small is an artist and professor in the School of Arts at California State University San Marcos. Her work focuses on documenting traditional and contemporary indigenous knowledge and uses of native plants in the southern California region. She does this through photography, writing, and an ethnobotany blog: deborahsmall.wordpress.com.

In 2010, she co-produced the *Contemporary Ethnobotanical Calendar* with Rose Ramirez, and then in 2015, they published the *Ethnobotany Project, Contemporary Uses of Native Plants, Southern California and Northern Baja Indians*. In 2013, she worked collaboratively with the **Yawáywish Program** hosted by the La Jolla Band of Luiseño Indians and others to produce two installations/exhibitions titled *Cultural Revitalization in San Diego County*.

Leslie Mouriquand

BIO

Leslie Mouriquand holds a BS in Geography and Anthropology from Cal Polytechnic University, Pomona, with an emphasis on California archaeology and indigenous cultures. Her MA was earned from California State University, Fullerton, in California Archaeology. Post graduate work was conducted at the University of Leicester, England, in ancient history and archaeology. Her research interests include traditional uses of plants for food and medicine, native plant propagation, and making handmade natural soaps. She is a certified **Master Desert Gardener.**

Leslie is a retired archaeologist for Riverside County, an adjunct professor of Anthropology at College of the Desert, and a cultural resources management/environmental planning consultant.

Cindi Moar Alvitre

BIO

Cindi Moar Alvitre *(Tongva)* has been an educator and artist-activist for over three decades. She served as the first woman chair of the *Gabrieleno/Tongva* Tribal Council as well as co-founding the **Ti'at Society,** an organization focused on the renewal of the ancient maritime practices of the coastal/island *Tongva.* In 1985, Cindi and Lorene Sisquoc co-founded **Mother Earth Clan,** a collective of Native women who created a model for cultural and environmental education, with a particular focus on traditional art.

Cindi is currently a PhD candidate at UCLA, Department of World Arts and Culture, with a focus on traditional medicine, revitalization, and cultural identity and trauma. As a cultural curator, her work extends beyond the physical manifestation of museum exhibition and into ceremonial performativity, a genre of expression that brings participants into native landscapes by "refocusing their cultural lens."

Craig Torres

BIO

Craig Torres *(Tongva)* is a member of the **Traditional Council of Pimu** and involved with the **Ti'at Society,** an organization focused on the revival of the traditional maritime culture of the southern California coastal region and southern Channel Islands. He is an artist, as well as *Tongva* cultural educator, presenter, and consultant for schools, culture and nature centers, museums, and city, state, and government agencies.

Besides his involvment with the Chia Café Collective, he is also part of the grassroots organization, **Preserving Our Heritage.** Both organizations provide cooking demos and classes with California native plants and emphasize the importance of preserving native plants, habitats, and landscapes for future generations.

Lorene Sisquoc

BIO

Lorene Sisquoc *(Mountain Cahuilla/Fort Sill Apache)* is Curator of the Sherman Indian Museum, and has taught Native Plant Uses and Material Culture/Traditions at Sherman Indian High School, as well as basketweaving throughout southern California for many years. She is co-founder of **Mother Earth Clan,** gives cultural presentations throughout the region, is on the board of directors of the Malki Museum, and is a co-founder of the **Southern California Indian Basket Weavers Organization.**

In 1997, the city of Riverside honored her with the **Dr. Martin Luther King, Jr. Visionary Award** for community cultural awareness. Lorene also received the **Dorothy Ramon Learning Center Heritage Keepers Award,** among many others.

Abe Sanchez

BIO

Abe Sanchez *(Purepecha)* is actively involved in the revival and preservation of Indigenous arts and foods. Two of his specialties are southern California American Indian basket weaving and native foods. His goal is to promote the decolonization of our diets by cooking and consuming native California and southwest plant-based foods. He works with traditional scholars and cultural specialists to learn culinary methods and cultural practices that he combines with his years of research and experience. He is particularly interested in traditional foods that are sustainable and readily available, yet underused. He believes that teaching about these ancient foods and helping people learn ways to gather, prepare, and eat them again will make a significant difference in our health and the health of our environment.

Heidi Lucero

BIO

Heidi Lucero *(Acjachemem/ Ohlone)* served as a member-at-large on the Acjachmen tribal council, working to promote cultural awareness. She currently holds a position on the NAGPRA committee at CSULB, facilitating the reburials of ancestoral remains, repatriation of tribal artifacts, and the housing and archiving of cultural items.

Heidi is a basket weaver, an accomplished traditional-style jewelry artist, and a member of the **California Indian Basketweavers Association.** She holds a BS in Anthropology from CSULB and plans to pursue her Masters in Cultural Sustainability. She also maintains relationship with Federal agencies with the goal of protecting and preserving California Native gathering places.

Tima Lotah Link

BIO

Tima Lotah Link *(Shmuwich Chumash)* is an advertising Art Director, cultural educator, and traditional textile artist. Her focus is on connecting people to the landscape by immersing them in the knowledge of traditional California Native cultures. Tima designs, lectures, and teaches for tribal organizations, educational institutions, non-profits, and governmental agencies. She designs the quarterly magazine, *News From Native California,* as well as cultural immersion experiences such as the *Living Cultures Chumash Exhibit* at the Bacara Resort & Spa and the *Native Garden* exhibit at the Autry Museum of the American West.

When she's not designing, her hands are busy weaving baskets featured in the book *California Indian Baskets: San Diego to Santa Barbara and Beyond to the San Joaquin Valley, Mountains and Deserts (Vol. II).* She received the **2012 Best in Show** at the Autry American Indian Arts Marketplace.

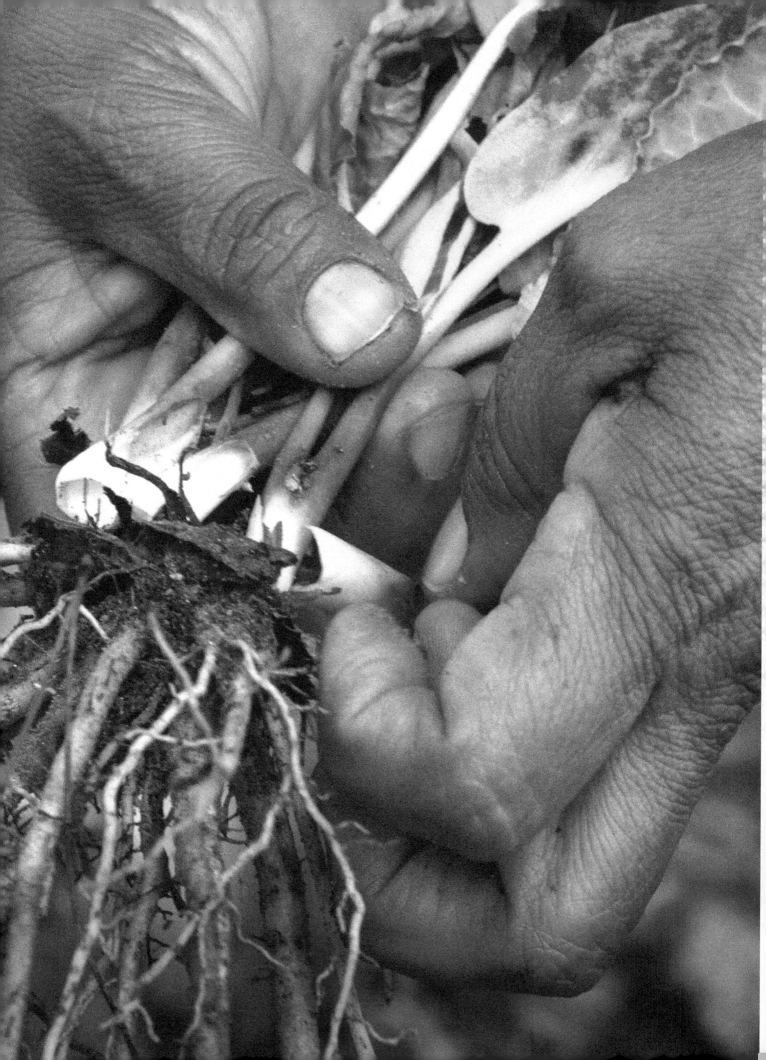

Resources

BOOKS

All Our Relations,
Winona LaDuke

American Indian Food and Lore,
Carolyn Niethammer

Braiding Sweetgrass,
Robin Wall Kimmerer

California Herbal Remedies,
Lolo Westrich

Chumash Ethnobotany,
Jan Timbrook

Cooking the Wild Southwest,
Carolyn Niethammer

Decolonize Your Diet,
Luz Calvo & Catriona Rueda Esquibel

EAT Mesquite!,
Desert Harvesters

Ethnobotany Project,
Rose Ramirez and Deborah Small

Feeding the People, Feeding the Spirit,
Elise Krohn & Valerie Segrest

Foods of the Southwest Indian Nations,
Lois Ellen Frank

Food Plants of the Sonoran Desert,
Wendy C. Hodgson

Gardening with a Wild Heart,
Judith Larner Lowry

Growing Food in a Hotter, Drier Land,
Gary Nabhan

In Defense of Food: An Eater's Manifesto,
Michael Pollan

Kumeyaay Ethnobotany,
Michael Wilken, forthcoming

The Landscaping Ideas of Jays,
Judith Larner Lowry

Medicinal Plants of the American Southwest,
Charles W. Kane

Medicinal Plants Used by Native American Tribes in Southern California,
Donna Largo, Daniel McCarthy, and Marcia Roper

Original Instructions: Indigenous Teachings for a Sustainable Future,
Melissa K. Nelson

Planting the Future,
Rosemary Gladstar

Rainwater Harvesting for Drylands and Beyond,
Brad Lancaster, Volumes 1 & 2

Recovering our Ancestors' Gardens,
Devon Abbott Mihesuah

Recovering the Sacred
Winona LaDuke

Seaweed, Salmon, and Manzanita Cider: A California Indian Feast,
Margaret Dubin & Sylvia Ross

Tending the Wild,
M. Kat Anderson

The Wild Wisdom of Weeds,
Katrina Blair

FEATURED BOOKS

Mulu'wetam, The First People,
Rosinda Nolasquez and Jane Hill

Temalpakh: Cahuilla Indian Knowledge and Usage of Plants,
Katherine Siva Saubel and Lowell John Bean

Yumáyk Yumáyk, Long Ago,
Villiana Calac Hyde and Eric Elliott

CALIFORNIA NATIVE PLANT NURSERIES

Back to Natives Restoration
www.backtonatives.org

California Native Plant Society: a listing of places to purchase native plants:
www.cnps.org/cnps/grownative/where_to_buy.php

Lily Rock Native Gardens Nursery
www.lilyrocknativegardens.com

Moosa Creek Nursery
www.moosacreeknursery.com

Rancho Santa Ana Botanic Garden (Grow Native Nursery)
www.rsabg.org

Theodore Payne Foundation Nursery
www.theodorepayne.org

Tree of Life Nursery
www.californianativeplants.com

ORGANIZATIONS

Agua Caliente Cultural Museum
www.accmuseum.org

Desert Harvesters
www.desertharvesters.org

Dorothy Ramon Learning Center
www.dorothyramon.org

Heyday Books
www.heydaybooks.com

Idyllwild Arts Summer Program
www.idyllwildarts.org

Malki Museum
www.malkimuseum.org

Native Seed/SEARCH
www.nativeseeds.org

Rancho Los Alamitos Historic Ranch and Gardens
www.rancholosalamitos.com

Rancho Santa Ana Botanic Garden
www.rsabg.org

Seri Women's Cooperative
Cooperativa Mujeres Productoras de Desemboque de los Seris
https://www.fondazioneslowfood.com/en/slow-food-presidia/seri-fire-roasted-mesquite/

Sherman Indian High School
www.shermanindian.org

Sherman Indian Museum
www.shermanindianmuseum.org

Tohono O'Odham Community Action (TOCA)
www.tocaonline.org

Torres Martinez Tribal TANF
www.torresmartinez.org

Acknowledgements

Lowell Bean, anthropologist, co-author *Temalpakh*

Tangie Bogner, *Cahuilla*, basketweaver

Sean Bogner, *Cahuilla*, basketweaver

Richard Bugbee, *Luiseño*, educator

Martha Burgess, Flor de Mayo Arts, ethnobotanist, watercolorist

Villiana Hyde Calac, *Luiseño*, educator, co-author *Yumáyk Yumáyk*, (1903-1994)

Gerald Clark, *Cahuilla*, educator

Heather Companiott, Director, Idyllwild Summer Arts and Native American Program

Teresa Castro, *Paipai/Ko'alh*, cultural educator

Diane L. Caudell, *Luiseño*, CIBA, basketweaver

Nephi Craig, *White Mountain Apache/Navajo*, chef

Teodora Cuero, *Kumiai*, elder, plant specialist, Traditional Authority of La Huerta, (1920-2014)

Valerie Dobesh, Native American Resource Center, master herbalist

Jane Dumas, Kumeyaay, educator, (1924-2014)

Bryan Endress, biologist, professor, Oregon State University

Mike Evans, Tree of Life Nursery, San Juan Capistrano

Naomi Fraga, Director of Conservation Programs, Rancho Santa Ana Botanic Garden

Lois Ellen Frank, *Kiowa*, culinary anthropologist, chef/owner Red Mesa Cuisine

Adriana Garcia, *K'iche Maya Kän/Xibalba Clan*, chef, caterer

Mike Gray, Quaker Workcamps, Pine Ridge and Desemboque

Nicholas Hernandez, *Cahuilla*, native plant landscaper, nursery manager at Arroyo Seco Foundation's Hahamongna Nursery

Kat High, *Hupa*, founder of *Haramokngna* American Indian Cultural Center

Terrol Dew Johnson, co-founder Tohono O'odham Community Action, TOCA

Monica Madrigal, *Luiseño*, educator

Malcolm Margolin, founder, *News from Native California*

Cadie McCarthy, caterer, chef

Sean Milanovich, *Cahuilla*, educator

Joseph Moreno, *Luiseño*, educator

Becky Munoa, *Luiseño*, educator

Pat Murkland, editor Ushkana Press, teacher, artist, writer

Joe Parker, International and Intercultural Studies, Pitzer College

Maren Peterson, Wallowa Resources, Communications and Development Coordinator

William J. Pink, *Cupeño/Luiseño*, Cultural consultant

Jennifer Ramirez, *Miwok/Serrano*

Petee Ramirez, *Miwok*

Rose Ramirez, *Chumash*, basketweaver, photographer, teacher

Peter Rice, Malibu Creek State Park presenter

Amy Rouillard, educator, Cabrillo National Monument

Laura Russo, photo: Yucca Moth (pg 134) standingoutinmyfield.wordpress.com

Stan Rodriguez, *Kumeyaay*, Bird Singer

Daniel Salgado, *Cahuilla*, educator

Antonio Sanchez, former nursery production manager, Rancho Santa Ana Botanic Garden

Katherine Siva Saubel, *Cahuilla*, educator, tribal leader, co-author *Temalpakh*, (1920-2011)

Ernest H. Siva, *Cahuilla/Serrano*, President and founder of Dorothy Ramon Learning Center, historian, educator, musician

Jean-Richard Perez, photographer, videographer

Larry Smith, *Lumbee*, American Indian Studies and Department of Film and Electronic Arts, CSULB

Erich Steinman, Sociology, Pitzer College

Jan Timbrook, anthropologist, ethnobiologist, writer

Beatrice Torres, *Purepecha,* Chef

Marian Walkingstick, *Acjachemen,* Tushmalum Heleqatum, basketweaver, educator

Roy Wiersma, Luther Burbank Spineless Cactus Identification Project

CSU San Marcos

Community Engaged Scholarship Incentive Grant: partial funding

Linda Muse, School of Arts, budget manager/administrative coordinator

Joely Proudfit, *Luiseño,* Sociology/Political Science; Director of California Indian Culture and Sovereignty Center

Tishmall Turner, *Luiseño,* Tribal Liaison

Rita Esmeralda Naranjo Realpe, Sociology alumna

Mike Wilken, anthropologist and ethnobotanist

The Pimu Catalina Island Archaeological Field School (PCIAFS)

www.pimu.weebly.com

Wendy Giddens Teeter, Desireé Reneé Martinez, and **Karimah Kennedy-Richardson,** Co-Directors PCIAFS

Blog

deborahsmall.wordpress.com

Facebook

www.facebook.com/ChiaCafeCollective/

CPSIA information can be obtained
at www.ICGtesting.com
Printed in the USA
JSHW051404230323
39319JS00018B/645